KU-489-803

Classification
of the
Acridomorphoid Insects

FRANCIS WALKER

From an original pencil sketch done rapidly at E. E. Austen's table in 1890 by A. G. Butler in response to Austen's enquiry as to Walker's general appearance. On completing the sketch Butler remarked that it "was very like him" (Walker), though he wrote on the sketch "legs to be lengthened". (Note by E.E.A. 2, IX, 1926.)

UNIVERSITY LIBRARY LIVERPOOL

Classification
of the
Acridomorphoid Insects

V. M. DIRSH

E. W. CLASSEY LTD
Faringdon, Oxon.

E. W. CLASSEY LTD
PARK ROAD
FARINGDON, OXON

SBN 900848 82 0
© V. M. DIRSH 1974
First Published 1975

Printed in Great Britain in the City of Oxford
Produced by Oxprint Limited

Contents

550794

Introduction

This work is an attempt to classify the Acridomorphoid insects and to group them into a system based on all the currently available characters of difference and similarity.

The latest preliminary system of *Acridoidea* (Dirsh, 1961) is now obsolete due to an accumulation of new facts and material in the intervening years. The author is now trying to create a system which, at present, will be adequate for use in Acridology. This is not a final system because a final system cannot be designed as long as *Acridoidea* exist. Changes are inevitable and of course they will follow.

In arranging material in systematic order the present author takes the principle of hierarchical upgrading of the higher taxa. This principle, as the author is aware, will not be accepted by all acridologists, which is inevitable. There are as many methods of classification and attitudes towards the value of taxa as there are acridologists. But it must be noted that most acridologists are extremely reluctant to make any changes in the classification of higher taxonomic units even in cases when such changes are obviously necessary. Acridologists are using classifications of the last century which are not only obsolete, but are, in many cases, confusing. In other groups of insects entomologists are less orthodox, for example about 30,000 species of known *Heteroptera* at present are successfully divided into 54 families (of course the *Heteroptera* is more diverse order than *Acridomorphoidea*) and a tendency to hierarchical upgrading of the taxa exists. However, as the clear definition of these hierarchical units is still lacking, the confused state in this respect is understandable.

For obvious reasons the author cannot give a list of genera of every family and subfamily — it would increase this work to an impractical size. The type genus is always given and sometimes some, or even all, genera of a taxon are given to clarify its scope.

The most unsatisfactory point in this review is, that there are several genera with one or two species, which do not fit into any of the subfamilies described, such as *Bactrophora* Westwood, 1845; *Rhicnoderma* Gerstaecker, 1889; *Ixalidium* Gerstaecker, 1869; *Mazaea* Stal, 1876; *Barombia* Karsch, 1891, and several more. There is also no doubt that more genera will be found which fit nowhere.

The author has two alternatives, either to create a new subfamily for each of these genera or to leave them as genera of indefinite position until further material is accumulated which will allow a decision to be reached in this respect. The present author chose the second alternative.

Classification and nomenclature

The classification of *Acridomorphoidea* (formerly *Acridoidea*) has changed during its history from Linnaean to the present time and undoubtedly will change in the future. In particular many changes occured in the hierarchic taxonomy, and in the nomenclature of the taxonomic units above the generic level.

1

The latest hierarchic scale proposed by Mayr (1969) is as follows:
Kingdom
 Phylum
  Subphylum
   Superclass
    Class
     Cohort
      Superorder
       Order
        Suborder
         Superfamily
         Family
          Subfamily
          Tribe
           Genus
           Species

However, past and present taxonomists have not adhered completely to this or any other existing hierarchic scale. Practically every taxonomist uses or has used his own hierarchic system which may differ a great deal from that presented above.

Some taxonomists consider the *Orthoptera* as a superorder, some as an order, others considered the taxon as a suborder or as a superfamily. In every case, every subordinate taxonomic unit was upgraded or downgraded accordingly and was shifted one or two steps up or down the hierarchic ladder. Frequently the hierarchic level was dependent on the existing contemporary tendency in this respect or more commonly from the individual test of each particular taxonomist.

A few examples of the classificatory schemes in this century may illustrate the chaotic state in which the hierarchic aspect of the classification still remains.

The majority of acridologists of this century considered the *Acridoidea* as a suborder of the *Orthoptera*. (Jakobson and Bianki, 1904; Handlirsh, 1908; Chopard, 1920; Schroder, 1925; Uvarov, 1928; Weber, 1938; Dirsh, 1961.) Some authors considered the *Acridoidea* as a superfamily (Ander, 1939; Chopard, 1949; Bey-Bienko & Mistshenko, 1951; Beier, 1955; Sharov, 1968; Imms, 1970) of the suborder *Caelifera* of the order *Orthoptera*. Some authors (Shvanwitsh, 1940) considered the *Acridoidea* as a superfamily, but omitted the suborder category and divided the order *Orthoptera* straight into superfamilies.

The chaotic state of hierarchic classification increased, disregarding existing categories and terms. The term "Section" was used as a taxonomic category between subclass and superorder (ignoring Cohort) (Kerkut, 1961). I. Bolivar, 1909 divided subfamily *Pyrgomorphinae* into *"Sections"*. Kevan (1970) divided the family *Pyrgomorphidae* into 30 tribes ignoring the subfamily category, but grouping the tribes into a category which he called "Series". No definition of this category was given and no diagnosis of these "Series" was offered.

The chaotic state of the hierarchic classification is probably a result of the absence of clear definitions of the hierarchic ranks of the system. Only a definition of the term "species" is formal and can be accepted as criteria of the rank. It is widely accepted now that a species is a certain population which differs from the other populations by characters of difference and cannot interbreed. All definitions of higher ranks are vague.

The genus is defined as a group of species (or one species) united by certain characters of similarity which are absent in other related genera.

The definition a family is as a "group of genera possessing characters which do not occur in other groups or some of which occur but in different combinations" (Dirsh, 1961). While a "subfamily" is regarded as a "group of genera with one or several convenient characters, which do not normally occur in other groups of the genera of a family, but are not exclusive, so that the occurence of intermediate genera is possible" (Dirsh, 1961).

Definitions of other ranks do not exist or used in vague terms of a group different from other groups by the characters of difference. This ambiguous definition may be applied to any rank and allows very wide interpretations.

In this work, which is concerned with the groups of family and subfamily ranks, the above cited definitions of these taxa are used. Also the following hierarchical divisions are used: the superorder, order, superfamily, family, and subfamily. Some contain tribes and certain groups are upgraded to the subfamily level, and the families are regrouped into superfamilies. The rank of suborder is omitted.

Ornithologists are in a most happy position. Almost all existing species of birds are described and fossil remnants are, if not abundant, then at least adequate, to classify birds and arrange them into a certain system. Mammalogists are in a similarly happy position with almost all mammals described and with an abundance of fossil material. Ichthyologists may be considered in position of advantage as well, because though not all species of existing fishes are known, there are plenty of fossils.

Orthopterologists and particularly Acridologists in this respect are in a greatly disadvantageous position. By the calculation of the present author, only one-third of the existing species of *Acridomorphoidea* are described; two-thirds of the pieces of the jig-saw puzzle are missing. Fossil material is so scanty that it is almost non-existent. Besides this, in all earlier described species, only the external, and most obvious characters were used, the internal anatomical characters in all taxonomic works were ignored, as well as were all physiological characters. Even at present the position is not much improved. Of course, the external morphological characters are revaluated and many new characters are discovered, and the structure of the genitalia is in general use, but internal skeletal structures, the structure of nervous system, digestive apparatus, tracheal net, are mostly not considered for purposes of taxonomy.

In these circumstances each system can be only tentative and must be changed periodically with new relevant discoveries.

In the present work the author has tried to use all characters which can be applied for classification. However, the external morphological characters and anatomical characters of genital structures are most used as the knowledge of other characters is too fragmentary.

The arrangement of the taxa of *Acridomorphoidea* used here is not a traditional one, and the present author is fully aware that many contemporary acridologists will disagree with it. However, it is necessary to point out that most orthopterologists are notoriously shy in rearranging and changing the existing system. They still adhere to the system and classification of the last century which was mostly created by Brunner von Wattenwyl. Since then hundreds of genera have been described and many new characters introduced. However, the new genera were placed into the same old system of groups only on the basis of superficial similarity, while in reality many of them are essentially different.

This has created heterogeneous higher taxa and an obsolete system, which is applicable with some degree of certainty only at specific and partly generic levels.

The nomenclature of Orthoptera has also undergone numerous modifications. Linnaeus in his "Systema Naturae" placed the group into the Order *Coleoptera*. Thunberg (1775) considered them as *Hemiptera*. Latreille (1793) erected a separate order — *Orthoptera* into which the *Acridoidea* and other unrelated groups were placed. In 1817 Latreille divided the Order *Orthoptera* into *Orthoptera-Saltatoria* i.e.

the jumping *Orthoptera* and *Orthoptera-Cursoria* — walking *Orthoptera*. As at that time the present orders *Dictyoptera, Dermaptera, Phasmoidea,* etc. were also placed in the *Orthoptera,* this division was justifiable.

Since then the order *Orthoptera* was named as such or as the order *Saltatoria,* sometimes the double named *Orthoptera-Saltatoria* was used. As late as this century and even in the present time the name *Saltatoria* is frequently used (Schroder, 1925; Weber, 1938; Ander, 1939; Imms, 1970).

Beier (1955) introduced a hybrid name *Saltatoptera* which was not accepted by Orthopterologists and was not used except by himself.

Ander (1939) divided the Order *Orthoptera* into two suborders and introduced two new names in the nomenclature — *Ensifera* into which he placed long-horned *Orthoptera,* and *Caelifera* into which the short-horned *Orthoptera* were placed. The division was artificial and unsatisfactory from a morphological point of view and from the point of view of the phylogenetic interrelationships of the groups. However, this division was used and is sometimes used even now.

It was felt by entomologists that the Order *Orthoptera* was too complex and contained diverse groups of various taxonomic value and not closely related. As a logical result of this general dissatisfaction the taxon Superorder *Orthopteroidea* was introduced.

The most recent and detailed division of *Orthopteriodea* was offered by Bey-Bienko (1962):

Superorder	*Orthopteroidea*
Orders:	*Blattoptera*
	Mantoptera
	Isoptera
	Plecoptera
	Embioptera
	Grylloblattida (=Notoptera)
	Phasmoptera
	Orthoptera
	Dermaptera
	Hemimerida
	Zoraptera

This division however is unsatisfactory in many respects, particularly in respect of the main differentiating characters and the omission of a character of great importance, the structure of genitalia.

Dirsh (1973) made the following re-arrangement and nomenclatural changes in the orthopteroid insect based mostly on morthological characters of difference and of similarity.

Superorder	*ORTHOPTEROIDEA*
Orders:	1. *Tettigonioidea*
	2. *Gryllacridoidea*
	3. *Grylloidea*
	4. *Gryllotalpoidea*
	5. *Cylindrochetoidea*
	6. *Tridactyloidea*
	7. *Rhipipterygoidea*
	8. *Tetrigoidea*
	9. *Eumastacoidea*
	10. *Acridomorphoidea*

The expulsion of the *Tetrigoidea* and placing them into separate order has been discussed already (Dirsh, 1961) and, it seems it is accepted now by the majority of the orthopterologists.

The consideration of eumastacids as a separate order is a rather new issue. Dirsh (1966) first erected the suborder *Acridomorpha* of the order *Orthoptera* and divided it into four superfamilies — *Eumastacoidea, Trigonopterygoidea, Pneumoroidea* and *Acridoidea* and used this division in his works 1968 and 1970. In 1973 Dirsh elevated eumastacids to the Order rank, while other superfamilies of *Acridomorpha* of the group were placed into a separate order — *Acridomorphoidea*.

In the present paper the following arrangement is proposed:

Superorder	*ORTHOPTEROIDEA*
Order	*EUMASTACOIDEA*
Superfamily	*Eumastaciidea*
Superfamily	*Proscopioidea*
Superfamily	*Trigonopteroidea*
Order	*ACRIDOMORPHOIDEA*
Superfamily	*Pneumoroidea*
Superfamily	*Pamphagoidea*
Superfamily	*Acridoidea*

Main characters used in classification

Theoretically all characters of the organism, both qualitative and quantitative can be used for classification. Characters at the atomic level — the kind and number of atoms of which an organism is composed; at the molecular level as a further step; the energetic level, considering the living systems of an organism as an energetic entity or the sum of metabolic processes. Cytology and histology may be used for the same purpose, and finally the form of an organism from the anatomo-morphological point of view (Dirsh, 1974).

In the present work the latter characters are mostly used. This is a complex of characters having certain limitations, but the author is completed to use them, owing to scanty and fragmentary knowledge of the other characters of the group.

Shape of body

This is one of the most important characters, as it gives the first impression of an insect. It was the first character used by all pre-Linnaean and post-Linnaean authors of the eighteenth and nineteenth centuries.

If used as a single character it is extremely deceptive. For example, from the time of Linnaeus and Fabricius and at the beginning of the nineteenth century the genera *Acrida* and *Truxalis* were considered as the same genus owing to their similar body shape and general appearance. However, they belong to different subfamilies. The unrelated genus *Pyrgomorpha* belonging to a different family, but with the same body appearance as *Truxalis* and *Acrida* was first considered as belonging to the genus *Truxalis* or *Acrida* Gmelin, 1790; Latreille, 1804; Lamarck, 1817; up to 1853, when Fischer first used *Pyrgomorpha* as a generic name (it was however used by Serville as subgeneric name in 1838 — *Truxalis* (*Pyrgomorpha*).

Only when other characters were introduced and used to supplement general appearance, entomologists began to differentiate between superficially similar genera.

There are many unrelated genera of different families and subfamilies which have a superficial appearance similiar to *Acrida, Truxalis* and *Pyrgomorpha* such as *Legua* Walker, 1870 of *Romaleinae* and *Mesopsera* of the *Hemiacrididae* which were misclassified.

As a further example of the general appearance deceiving an author to place unrelated genera of different families into one taxon. Uvarov (1943) placed into tribe *Phrynotettigini* of the *Pamphagidae* (Sensu Uvarov) the following genera: *Phrynotettis* Uhler, 1872; *Dracotettix* Bruner, 1889; both of *Romaleinae, Buforania* Sjostedt, 1920 (of the *Catantopinae*); *Aucacris* Hebard, 1929 (of *Aucacrinae*). There is nothing in common between them except a certain similarity in general appearance.

Whether the similarity of appearance in body form is the result of parallel development in different lineages, or it is convergence resulting from adaptation to the same ecological niche and developed because of natural selection, cannot be said at present. Probably a combination of several factors ought to be taken into account. At any rate the great diversity in body form and also great similarity can be observed and used as a feature supplementary to other characters.

It ought to be mentioned also that body length in *Acridomorphoidea* varies from 5 mm. (in *Illapelinae*) to 101 mm. (in *Truxalinae*). The ratio of length to width of body (index of body proportion, Dirsh, 1968) varies from 2 in *Lathicerus cimex* to 20 in *Cannula linearis*.

The shape of the body of the imago can be classified as follows:
    a. Cylindrical, or subcylindrical, this being the most usual shape.
    b. Narrowly cylindrical, usually described as stick-like (for example, *Acrida,* Truxalis).
    c. Very narrowly cylindrical, described as straw-like (for example, *Cannula, Proscopia, Mesopsera*).
    d. Laterally compressed, so that height greater than width (for example, *Plagiotriptus,* some *Tropidauchen, Xyronotus*).
    e. Dorso-ventrally compressed, so that width greater than height (for example, *Lathicerus, Trachypetrella, Batrachotetrix*).

Head

The shape of the head cannot be used as a single character for diagnosing every family or subfamily, owing to its variability, but for some of them such as *Acridinae, Truxalinae, Opshomalinae, Leptisminae* it is quite characteristic.

The presence or absence of the furrow of fastigium of vertex is the most important character of the head. It has been regarded as a primary epicranial suture which has persisted in some families and subfamilies in the imago, but has disappeared in others. However, the latest opinion (Du Porte, 1946; Snodgrass, 1947) is that the so-called epicranial suture is no suture at all. It does not separate the sclerites, and it shows no indication of the place of their fusing. Matsuda, 1965, considered that this "suture" is the ecdysial cleavage line, along which the cuticle is weakened for effecting rupture at ecdysis.

The fastigial furrow is always present in the imago and in all the nymphal instars in the following taxa: *Pneumoridae, Trigonopterygidae, Xyronotidae, Charilaidae, Pamphagidae, Lathiceridae, Pyrgomorphidae, Ommexechidae*.

The meaning of the fastigial furrow from an anatomical point of view is unknown. Possibly it is a skeletal development for strengthening the structure of the cranial capsule. To support this view, it is noteworthy that even in *Pamphagidae* it is more developed in the species with the head conical than in those with the head subglobular.

This character is used for purely phenetic classification, its phylogenetic meaning being unknown. Speculations on its phylogeny cannot be profitable until its development in relation to the internal structure of the cranial capsule has been studied. It should be noted though that the fastigial furrow is present in more primitive and absent in more advanced families and subfamilies.

Antennae vary in the shape, in length and number of segments in almost every subfamily. However the shape of antennae, is a character which has a diagnostic value in certain subfamilies. The number of segments may vary from 8 to 33 and in females the number of segments is usually larger than in males (Mason, 1954).

The presence or absence of fastigial foveolae or analogous structures; is used widely in grouping taxa and is an important character for diagnosing subfamilies and taxa of lower rank. The same is applicable to fastigial areolae in the family *Pyrgomorphidae*.

The meaning and function of these structures is unknown, but probably they are places of location of certain sense organs.

Antennal grooves only occur in the family *Lathiceridae*. Their functional meaning is clear — to fold into them antenna when insect burrow through a soil.

Thorax

The shape of the thorax and its sculpture particularly the pronotum is extremely variable in almost every subfamily. As a diagnostic character at the subfamily level its usefulness is rather limited. However, in some subfamilies it has certain value in this respect.

The prosternal process, tubercle, elevation, or collar, is one of the most useful characters for diagnostic purposes in families and lower systematic taxa. The function of this anatomical entity is not known, but its presence or absence and its shape provide taxonomy of Acridomorphoidea with one of the most important features of some of the taxa.

The shape of the mesosternal lobes and mesosternal interspace is characteristic for certain subfamilies and for the *Cyrtacanthacrinae* and the *Egnatiinae* form of the mesosternal lobes is visually the only reliable character to differentiate these subfamilies.

Tegmina and wings

Tegmina and wings vary from fully developed, shortened, lobiform-lateral, vestigial to completely absent. All of these variations are characters of utmost importance. There are families and subfamilies fully apterous, but in all probability the absence of organs of flight is not a primary character, but secondary loss of these organs.

The most important feature of the tegmina and wings is their venation. The character which is used in the diagnoses of families and subfamilies and sometimes, as in *Pneumoridae*, is the feature of winged family of the whole superfamily *Pneumoroidea*. Various authors in the history of the classification of acridids interpreted the veins and areas between veins differently and sometimes confused them.

In 1955 Ragge presented a unified system of venation for all *Orthoptera* (in the sense of the scope of the taxon at that time), and since then his system has been adopted by almost all acridologists. In this work the system of venation is after Ragge (1955) with slightly modified terminology of the anterior part of the tegmen and wing which is called the remigium, and the posterior part which is called the vannus. The parts are separated by the vannal fold, the main fold along which the wings are folded (Snodgrass, 1935).

Another important character is the specialisation of the tegmina and wings for producing sounds. Sometimes it is a specialisation of main veins, sometimes it is a specialisation of veinlets in various areas of the tegmen and wing. Such specialisation is characteristic for certain subfamilies.

Tympanal organ

The tympanal organ (or tympanum) is the main character separating the orders *Eumastacoidea* and *Acridomorphoidea*. In the *Eumastacoidea* it is absent and probably never has been developed. In the *Acridomorphoidea* it is normally present but if it is absent, this is probably a secondary development in the evolution of the group.

According to Mason (1969) the structure of the tympanal organ is characteristic for the families of *Acridomorphoidea* and can be used as a supplementary character.

Legs

In the classification of *Acridomorphoidea* only the hind legs have important diagnostic characters for taxa down to the subfamily level. One of the most important diagnostic characters is the comparative length of the upper and lower basal lobes of the hind femur. On the basis of this character the families and subfamilies are differentiated.

The second diagnostic character is the structure of the lower lobe of the hind knee, which may be rounded, angular, or spine-like. In the *Oxyinae*, it is a decisive character for the differentiation of this subfamily.

The inner side of the hind femur frequently bears a ridge, serration, or row of articulated pegs as part of a sound producing mechanism. This character has diagnostic value for certain subfamilies.

The shape of the hind femur has a certain value as a diagnostic character as well. The ratio of length to width varies between 2 and 21, and the general shape varies to a great extent.

The general shape of the hind tibia and particularly the presence or absence of the external apical spine are characters, particularly the latter one, of great value.

Sound producing mechanism

There are many types of sound producing mechanism. If the mechanism has not been found, most probably it is difficult to find it as a morphological structure. The present author believes that all or almost all acridids possess some kind of mechanism to produce sound. Sound is produced by friction of the hind femora against the tegmina, or friction of the hind femur or tibia against abdominal ridges, or friction of the tegmina against the pronotum, and numerous other mechanisms. Some of them are undetectable even if a sound is audible. The *Proscopiidae* produce a faint squeak, probably by the emitting air from their gut. This latter sound was detected by the present author and his assistant Miss J. B. Mason while the actual mechanism was not found. Some sounds produced by a well developed mechanism are probably ultrasonic; in the morphological sense the sound producing mechanism is unmistakable, but the sound is not detected by human ear as in *Tanaocerus*. Probably in morphologically less detectable mechanisms ultrasonic sounds are produced as well.

At present, for classification purposes, only morphological mechanisms for sound production are used if they are clearly detectable The sound itself may provide a good supplementary character, but it is not studied yet to such an extent that it can be used for purposes of classification of taxa higher than genus.

In this work only two types of sound producing mechanism are used — the tegmino-femoral mechanism and abdomino-femoral mechanism.

Genital organs

Male: External secondary analo-genital organs, such as the supra-anal plate, cerci, subgenital plate do not provide characters for the differentiation of taxa above the generic level, except in a few subfamilies — *Calliptaminae* and *Euryphyminae*.

The main taxonomic character in the classification of *Acridomorphoidea* however, is the morphological structure of the inner genitalia, particularly the phallic complex in males and partly the structure of the spermatheca in females. The phallic complex, at present, is extensively studied, to the extent that, on the basis of this character alone the acridids can be separated into taxa of subfamily level and higher taxa.

The phallic complex primarily represents the widened distal end of the ejaculatory duct. The internal part of this widening is the endophallus. The external part, formed as an external evagination of the endophallus, is the ectophallus. The main part of the endophallus is the endophallic sac, which is a simple but widened continuation of the ejaculatory duct, and is divided into the ejaculatory sac, spermatophore sac and phallotreme. Sometimes this division is not clear, particularly in more primitive families and subfamilies, but in most cases is quite definite. From the functional point of view, the ejaculatory sac represents a reservoir for storing sperm and spermal fluid. The spermatophore sac represents part of the endophallic sac where the spermatophore is formed ready for transplanting into the female genital chamber. Usually both sacs are connected by the gonopore but sometimes, in primitive families, they are simply continuous. The spermatophore is transmitted through the phallotreme and its sclerotized ends, or with the help of the penis, into the female genital chamber. This transmission is affected by the active movement of the skeletal parts of the endophallus and partly the ectophallus (valves of cingulum). The active movement of the skeletal parts is achieved by the strong muscles attached to them (they are poorly studied, because the study of genitalia has been mostly performed on dry material). The second factor, facilitating the movement of the spermatophore, is strong blood pressure which makes the whole phallic complex erect and swollen. A combination of these factors makes possible the penetration of the whole distal part of the phallic complex into the genital chamber of the female and movement of the spermatophore. Further movement of the spermatophore may be effected by the active movement of spermatodesms as suggested by Pickfor and Gillot (1971) or by chemotaxis in the spermathecal duct, or by the sucking forces of the spermatheca itself, which is covered by a layer muscle, and probably can be squeezed and relaxed making pumping movements. Which of these factors is most important in this process is not clear and it is possible that all are involved.

The skeletal parts of the endophallus forming the penis are most important functionally, and for taxonomic purposes, are most important morphologically. They are probably derived from the walls of the endophallic sac and represented by a pair of sclerites of various shape. Their function is to regulate and direct the flow of the sperm from ejaculatory duct to the ejaculatory sac and from the ejaculatory sac to the spermatophe sac. Their function is also to penetrate into the female genital chamber and into the spermathecal duct to facilitate the transfer of spermatophores. In primitive families the penis' sclerites are hardly detectable or absent.

The pair of strongly sclerotised endophallic sclerites are present in all advanced families of *Acridomorphoidea* but in *Eumastacoidea* the penis is monoscleritic.

The structure of the penis sclerites in *Acridomorphoidea* is a most important diagnostic character. The sclerite may be represented by a single piece, or it may be divided into basal and apical valves fully separated and connected only by the walls of the endophallic sac, or separated but connected by a sclerotized flexure. In seven families the penis' sclerite has a gonopore processes which is a part of the mechanism regulating the movement of sperm from the ejaculatory to the spermatophore sac. In families in which the gonopore process is absent the flow of sperm is probably regulated by the penis' sclerites.

The shape of the penis sclerites is the most important diagnostic character, at all levels of the taxonomic hierarchy.

The position of the sacs of the endophallus is also an important diagnostic character. The spermatophore sac is in a dorsal position in the families in which the gonopore processes are absent and in the middle or ventral position in most cases when the gonopore processes are present.

The ectophallus probably plays a less active role in the movement of the phallic complex, and its function is mostly to support the whole phallic system. However the valves of the cingulum take an active part in transferring the spermatophore together with the apical valves or apical part of the penis into the female genital chamber. The cingulum, with all its parts as well as all other ectophallic sclerotisations, is probably derived from the ectophallic membrane. Differentiation of the cingulum or its absence is also a very important diagnostic character.

The epiphallus is a rather separate organ. It is derived from the ectophallic membrane and often is so closely connected with it as to be considered as a sclerotized part of the membrane. Its function is supplementary, and in the act of copulation it stabilises the whole phallus and firmly keeps the position to allow the distal part of penis to penetrate into the female genital chamber. The epiphallus occurs in all *Acridomorphoidea* and has various shapes and characteristics for every taxon of the order and diagnostic value. Auxillary parts of the epiphallus, the oval sclerites and lateral appendages are probably sclerotized parts for the attachment of powerful muscles, which move the epiphallus and hold it in certain positions.

Female: External secondary genital organs in the female provide many important diagnostic characters for the classification of *Acridomorphoidea* below the subfamily level and rarely above subfamily level. The supra-anal plate and cerci are rather uniform in all systematic units. The subgenital plate, its shape and structure on its ventral and dorsal surfaces, provides characters which can be used in the differentiation of one subfamily — the *Oxyinae,* but mostly are useful in taxa below subfamily rank. The ovipositor, strictly speaking, is not part of the genital mechanism, but an instrument for digging and depositing eggs. However, the ovipositor is characteristic for the family *Pauliniidae.*

The most important character of the female genitalia is the structure of the spermathecal system, which sometimes may be characteristic for various taxa. Regretfully it has not been studied enough to be useful for classification as a decisive character. Slifer (1939, 1940, 1943) who studied most extensively the structure of spermatheca established a terminology for its parts — spermathecal duct, apical and preapical diverticula and secondary diverticula. The present author found that the terms apical and preapical diverticulum rather confusing, because it is not always possible to decide which is the apical diverticulum and which is the preapical diverticulum. Therefore, the terms 'main reservoir' and 'diverticulum' are used instead.

The genitalia of *Acridomorphoidea* and *Eumastacoidea* are considered here as the main character for differentiation of the higher taxa. During process of evolution all characters were adapting to certain ecological conditions, which were changing drastically in geological time. Adaptation to different food, appearing in certain ecological niche, to climatic condition, to inter-relation with other groups of animal etc. necessitating certain morpho-anatomical changes. The characters evolve by adapting to new demands, and the selective process brought the new forms.

The genitalia during evolutionary history, particularly internal genitalia, served a single and invariable task — fertilisation. Their changes during evolution were more orthogenetic than changes of other organs and structures of a body. Their evolution, of course, was influenced by the co-ordinative changes of the whole organism and its biological system, but the invariability of function and their internal location led to more stability of their structure and to less flexibility towards external environmental conditions. They are more orthogenetic in their development, more conservative, and more readily reflect the ancestral structures of the organs.

Karyotype

Since the chromosomes were discovered, they have attracted attention in all branches of biology. Taxonomists of every group of plants and animals also took interest in them as a means for classification and as a means to find interrelationships between taxa. Acridology has had its share in this respect.

As early as 1908 McClung, referring to Acridids expressed an idea that chromosomes show specific, generic and family characters as a whole organism. In 1917 McClung, using the shape of chromosomes separated *Mermiria bivittata* into two groups. This division was confirmed later by Rehn (1919) who considered these two groups as separate subspecies.

The application of karyotype characters since then has become quite common.

From the beginning of the present century the question of numbers and the morphology of chromosomes and the interpretation of these phenomena has been greatly argued. This makes the application of karyotype characters for taxonomic purposes difficult and uncertain, but the main trend is still very indicative and can be used in taxonomy.

White (1957, 1970) stated that the number of chromosomes in acridids varies from 23 to 8. In his opinion the basic number is 23 and reduction is the result of fusion and inversing of the chromosomes. Helwig (1958) is however of the opinion that the more primitive families (*Pyrgomorphidae* and *Pamphagidae*) have 19 chromosomes and the other, more advanced, families with 23 chromosomes acquired the additional chromosomes during the process of evolution. The controversy continues and the cytological mechanism of the process is still not solved either (John & Hewitt, 1968; White, 1969). For purely taxonomic purposes the mechanism of the fusion of the chromosomes and acquiring the super numerary chromosomes is not considered of primary importance in this paper. Most important for taxonomists is the final result. However, when using karyotype characters these points must be taken into consideration.

The first application of karyotype characters to the higher taxonomic units in acridids was offered by Helwig (1958). First of all the families were divided according to the size of the chromosomes:- small size, medium size and large size.

According to Helwig the families in the group with small sized chromosomes are:- *Pneumoridae, Xyronotidae, Trigonopterygidae, Proscopiidae, Ommexechidae, Pauliniidae, Lentulidae,* and all *Eumastacidae* (they were considered at that time as a family of *Acridoidea*).

The family *Acrididae* belongs to the group of medium sized chromosomes. The group large sized chromosomes consists of the families *Pyrgomorphidae* and *Pamphagidae*.

The small sized chromosome karyotype was considered by Helwig as more primitive than other karyotypes. He based this suggestion on the fact that the genera *Tridactylus* and *Rhipipteryx*, which according to Crampton (1927) "occupy a position near the base of the lines of descent of the Acridoid families", also have a small-sized chromosome karyotype, and are more primitive than the recent *Acridoidea*. The latter, during the process of evolution, acquired additional chromosomes, but the size of some of them reflects the character of the ancestral stock.

I do not propose to discuss the phylogenetic speculation expressed above, but it is necessary to point out the following contradiction. From one side the *Pyrgomorphidae* and *Pamphagidae* have the largest size of chromosomes which Helwig considered as an advanced character but at the same time they have only 19 chromosomes, which he considered as a primitive character. It is necessary to note also that the grouping of the families of *Acridoidea* according to the size of chromosomes has not met with a wide acclamation since Helwig's initial work.

The number of chromosomes in the families investigated so far however is a character more readily acceptable. According to Helwig the following families possess 2n σ = 23 chromosomes — *Pneumoridae, Xyronotidae, Trigonopterygidae, Ommexechidae, Pauliniidae, Lentulidae* and *Acrididae*.

Pyrgomorphidae and *Pamphagidae* have 2n σ = 19 chromosomes. The *Eumastacidae* complex has 2n σ = 17, 19, 21 or 23 chromosomes. The family *Proscopiidae* has 2n σ = 17 chromosomes. The newly investigated family *Charilaidae* (White, 1967) has 2n σ = 23 chromosomes. The families *Tanaoceridae* and *Lathiceridae* have not been investigated in this respect.

The basic number of chromosomes is characteristic for a family, but the deviations owing to supernumerary or reduction of the number of chromosomes are quite frequent. This prevents the consideration of karyology as an absolute taxonomic criterion, but as a useful additional taxonomic character. The prevailing number of chromosomes in a family of subfamily is still characteristic, and with certain reservations can be used in taxonomy.

The present author suspects that the supposed polymorphic karyotype in some species is frequently due to misidentification.

Probably the karyological abnormalities in a population of *Chorthippus brunneus* (John, Lewis, Henderson, 1960) are due to the fact that the population was a mixture of infraspecific taxa, and was not investigated thoroughly in this respect.

However, the fact cannot be overlooked that the prevailing number of chromosomes in the subfamily *Truxalinae* is 2n σ = 17, while in other subfamilies of *Acrididae* 2n σ = 23.

At any rate, at present the cytologist has not created any system or even a crude classification of *Acridomorphoidea* on the karyotype basis, and when they study chromosomes of a certain species they always ask an orthodox morphological taxonomist, — what species is it?

SUPERORDER
Orthopteroidea

Diagnosis: Hemimetabolic insects. Mandibulate. Male internal genital organs symmetrical. Cerci present. Two pairs of wings mostly present, fore-wings specialised as cover for hind wings. Fertilisation by spermatophores, directly inserted into spermathecal duct.

Orthopteroidea contains the following orders:

1. *Tettigonioidea*
2. *Gryllacridoidea*
3. *Grylloidea*
4. *Gryllotalpoidea*
5. *Cylindrochetoidea*
6. *Tridactyloidea*
7. *Rhipipterygoidea*
8. *Tetrigoidea*
9. *Eumastacoidea*
10. *Acridomorphoidea*

The main difference between the superorder *Orthopteroidea* and other more closely related superorders of insects is that the male internal genitalia are symmetrical while in the orders formerly placed into *Orthopteroidea* — *Blattoidea, Mantoidea, Grylloblattoidea, Phasmoidea, Embioidea* and *Dermaptera,* they are asymmetrical.

ORDER
Eumastacoidea

Body of variable shape. Tympanal organ primarily absent. Distal antennal organ present. Tegmina and wings fully developed, reduced, or absent. Brunner's organ present, (sometimes rudimentary). Dorsal surface of metatarsus of hind legs mostly armed.

Phallic complex highly specialized and structurally complicated. Penis or analogous organ, if present, primarily monoscleritic; epiphallus or analogous structure mostly present; oval sclerites absent.

Order *Eumastacoidea* contains the following superfamilies:

1. *Eumastaciidea*
2. *Locustopseidea* (fossil)
3. *Proscopiidea*
4. *Trigonopterygoidea*

The main characters differentiating the *Eumastacoidea* from the *Acridomorphoidea* are: Primarily absence of tympanum; mostly armed dorsal side of hind metatarsus, and monoscleritic penis (if present).

Key to superfamilies

1 (2) Ileal caeca present. Distal antennal organ well developed. Brunner's organ well developed. Spermatheca simple, ampoula-like, metatarsus of hind leg, mostly armed.

Eumastaciidea

2 (1) Ileal caeca absent. Distal antennal organ present or rudimentary. Spermatheca of various shape. Metatarsus of hind leg unarmed.

3 (4) Body straw-like, head elongated, narrow conical. Pronotum tubular. Penis monoscleritic. Mostly apterous or with strongly reduced wings. Phallic complex in normal position.

Proscopiidea

4 (3) Body subcylindrical. Head acutely conical. Pronotum rectangular (in cross section). Penis mono- or biscleritic. Fully winged or with wings shortened. Phallic complex in reversed position.

Trigonopterygoidea

The members of the order *Eumastacoidea* in externo-morphological appearance and in the function of their external appendices, are similar to members of the order *Acridomorphoidea*. This is probably due to the converging evolutionary development of both groups, prompted by the reason that both of them are phytophagous and were living in similar habitats. The phylogenitical affinities of these groups, however, are still rather obscure. It can be speculated that they were related as very ancient ancestral branches of insects, the nature of which we cannot even guess, owing to the absence of fossil evidence. Probably they originated from the common stock of the hypothetical *Protorthoptera*.

From the middle of the last century entomologists considered eumastacids as a family. Dirsh (1966) upgraded them to the rank of superfamily of the suborder *Acridomorpha*.

Sharov (1968) placed the *Eumastacidae* as a family into the superfamily *Locustopseidea,* together with two fossil families, the *Locustavidae* and *Locustopseidae* (in his hierarchical scale a superfamily is equivalent to an order in the present work), basing this arrangement entirely on wing venation.

The order *Eumastacoidea* is a highly heterogenious complex and probably is polyphyletic in its origin. It is composed of families, the phylogenetic relations of which are obscure and, in fact, cannot be properly assessed at present.

Only recently the order has begun to attract the attention of entomologists and fragmentary material is being accumulated in collections. One of the reasons for this inadequate knowledge of the group is their rarity, cryptic habits and the general difficulty of collecting them.

Since almost all morphological characters except venation in fossil insects are lost, comparison between living and fossil superfamilies is possible only on venation. According to Sharov (1968) the difference is that in *Eumastaciidea* the medial vein of tegmen has two branches or is unbranched, and there are no anastomoses between the medial and cubital veins. In the *Locustopseides* the medial vein contains no less than three branches, and if its posterior branch is fused with the cubital vein, then its base is preserved as an oblique vein.

Sharov proposed the existence of two systematically close fossil families which are regarded here as one superfamily, the *Locustopseidae* Handlirsch, 1906 and *Locustovidae* Sharov, 1968. The difference between these families, according to Sharov, 1968 is as follows:

In *Locustopseidae* the cubital anterior vein (CuA) is longer than CuA_2, and the latter branches from CuA near to the middle of the wing.

In *Locustavidae* the cubital anterior vein (CuA) is shorter than CuA_2, and the latter branches from CuA at the base of the wing.

The palaeontological history of the order is still not very well known owing to the lack of material.

Scudder (1890) described a fossil eumastacid from Miocene deposits, and Cockerell (1909, 1926) also described eumastacids from the Miocene. Handlirsch (1910) recorded a eumastacid from the Oligocene. Zeuner (1937, 1939, 1941, 1942, 1944) discovered eumastacids from the Miocene and Oligocene.

Handlirsch (1908) expressed the opinion that primitive Acridids evolved from Triassic and Liassic.

Ragge (1955), on the basis of a comparative study of the venation of Orthoptera and related groups of insects placed the roots of the origin of *Eumastacidae* and *Locustopseidae* into the middle Permian period.

Sharov (1968) described *Eumastacidae* from the Upper Jurassic and *Locustopseidae* from the Upper Cretaceous, Upper, middle and lower Jurassic, and the lower Triassic, and *Locustavidae* from lower Triassic deposits.

Blackith (1973) analysing fossil data from the point of view of ecological conditions and particularly feeding habits of present eumastacids supported the point of view that the roots of the origin of the group are in the end of the Palaeozoic era.

Not having adequate material to make definite conclusions the present author, after studying comparative morpho-anatomical characters of *Orthopteroidea* and related groups, has come to the conclusion that the ancestral forms of the *Eumastacoidea* came from the late part of the Carboniferous period and flourished in the Permian period.

Superfamily
Eumastaciidea

Diagnosis: Body compressed, cylindrical or cylindrically elongated. Head from vertical to horizontal; face flattened or incurved; costa frontalis from well developed to almost obliterated, sulcate or subsulcate. Basi-occipital slit present or absent; dorsal cervical membrane with or without sclerotized plates. Antennae filiform, clavate or ensiform. Antennal organ present. Prosternal process absent. Macropterous, brachypterous, micropterous or apterous. Hind femora from strongly compressed to normal type. Brunner's organ present. Dorsal surface of hind metatarsus concave and mostly armed.

PHALLIC COMPLEX: Ectophallus specialized; endophallus clearly differentiated. Penis mostly monoscleritic. Ejaculatory duct connected with ejaculatory sac in dorso-proximal part.

Spermatheca mostly consist of single reservoir.

The superfamily is divided into the following seven families:

1. *Chorotypidae*
2. *Eruciidae*
3. *Eumastacidae*
4. *Euschmidtiidae*
5. *Gomphomastacidae*
6. *Morabidae*
7. *Thericleidae*

These families are separated by the structure of the phallic complex and in all probability are of polyphyletic origin.

Geographically they occupy different areas, but sometimes these areas overlap.

Key to families

1 (12) Penis monoscleritic, or biscleritic, well developed.

2 (3) Penis monoscleritic, rod-like or filiform. Spermatophore sac absent or rudimentary.

Thericleidae

3 (2) Penis monoscleritic or biscleritic. Spermatophore sac well developed.

4 (11) Penis simple, U-shaped, monoscleritic, with prongs directed towards distal end.

5 (6) Ectophallus strongly specialized, with numerous teeth. Epiphallus not clearly differentiated from ectophallus (homology with epiphalli in other families is doubtful).

Eruciidae

6 (5) Ectophallus comparatively simple. Epiphallus well differentiated.

7 (10) Epiphallus mostly bridge-shaped, strongly sclerotized.

8 (9) Basi-occipital slit absent. Antennae filiform or clavate.

Gomophomastacidae

9 (8) Basi-occipital slit present. Antennae ensiform.

Morabidae

10 (7) Epiphallus shield-shaped, weakly sclerotized in middle part.

Eumastacidae

11 (4) Penis complicated, biscleritic or monoscleritic sometimes fused at distal end.

Euschmidtiidae

12 (1) Penis poorly developed monoscleritic or absent.

Chorotypidae

Family

Thericleidae

(Fig. 1)

Diagnosis: Body usually laterally compressed or subcylindrical. Head vertical or subvertical; face flattened; costa frontalis mostly narrow, well developed or obliterated in lower part, sulcate or subsulcate. Basi-occipital slit absent; dorsal cervical membrane without sclerotized plates. Antennae short, filiform, 9-12 segmented. Apterous or with reduced wings. Hind femora strongly or slightly compressed. Spurs on both sides of hind tibia calcariform; dorsal side of hind metatarsus with external apical tooth and sometimes with basal tubercle.

PHALLIC COMPLEX: Ectophallus relatively small, sclerotized, covering only apical part of endophallus, and highly specialized. Endophallus relatively large, clearly differentiated from ectophallus, and of highly complicated structure; ejaculatory sac small, in dorso-proximal position; ejaculatory duct connected with the sac in dorso-medial part; spermatophore sac absent or substituted by very small chamber in base of penis (fig. 1 *f*); phallotreme absent and substituted by narrow penis which consists of single, narrow tube, flexible along whole length except strongly sclerotized apical part. Epiphallus arch-shaped with pair of latero-apical, strongly sclerotized, hooks.

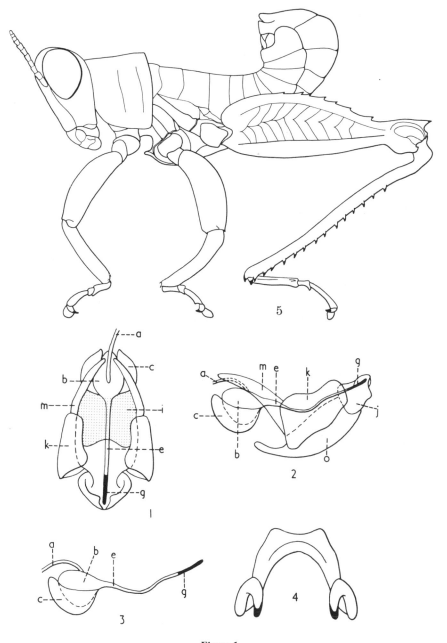

Figure 1.

5, *Thericles euchore* C. Bolivar, 1914. Male. 1-4, phallic complex of *Thericles whitei* Dirsh, 1964. 1, phallic complex from above, epiphallus removed. 2, the same, lateral view. 3, endophallus, lateral view. 4, epiphallus.

a, ejaculatory duct. b, ejaculatory sac. c, lateral "wing" of ejaculatory sac. e, soft, weakly sclerotised duct, merging with the apical, strongly sclerotised part (g). g, strongly sclerotised duct, representing continuation of the duct e and considered as the penis. i, internal envelope, covering apical part of endophallus. j, external envelope, covering apical part of endophallus. k, strongly sclerotised lateral plates of ectophallus. m, strongly sclerotised bar-like formation of ectophallus. o, ventral, membraneous part of ectophallus.

Spermatheca ampoula-like.
Karyotype: 2n ♂ = 17, 19, predominantly 2n ♂ = 19.
Type genus: *Thericles* Stal, 1875.

Thericleidae according to the structure of the phallic complex are distinctly isolated from the other families of the superfamily. Its structure, is sometimes comparatively simple, sometimes highly complicated, but the general plan is the same. The main characters being the monoscleritic rod-like penis and the rudimentary or absent spermatophore sac.

Distribution: Ethiopian Region including Arabia and Socotra and other small coastal islands of Africa.

Contains the subfamily *Thericleinae* Burr, 1903 and subfamily *Socotrellinae* Popov, 1959. Descamps (1970) rightly synonymised both subfamilies on the basis of similarity in the structure of the phallic complex.

Family

Eruciidae

(Fig. 2)

Diagnosis: Body subcylindrical. Head vertical; face moderately or not at all flattened; costa frontalis low, sulcate or subsulcate. Basi-occipital slit absent; dorsal cervical membrane without sclerotized plates. Antennae short, filiform, 12-13 segmented. Macropterous or with reduced wings. Hind femora not compressed; spurs on both sides of hind tibia normally developed, calcariform; dorsal side of hind metatarsus serrated or spined.

PHALLIC COMPLEX: Ectophallus strongly specialised, relatively very large, membraneous, capsule-like, at apex and on dorsal side with strongly sclerotized teeth. Endophallus clearly differentiated from ectophallus, relatively very small, with strongly sclerotized penis, forming U-shaped fork with prongs directed towards distal end. Endophallic sacs relatively small; ejaculatory sac in ventro-proximal position; ejaculatory duct connected with the sac in dorso-proximal part; spermatophore sac in ventro-distal position; phallotreme short and wide, opened between and below apical prongs of penis. Epiphallus not clearly differentiated, moderately sclerotized, forming shield-like, large structure with pair of large apical and lateral appendages.

Spermatheca ampoule-shaped, simple.
Karyotype: 2n ♂ = 21. (Based on a few observations).
Type genus: *Erucius* Stal, 1875.

This family was regarded by Burr (1903) as a subfamily. However, the structure of the phallic complex of the group differs so much from other eumastacids that the only possibility remaining is to place it into the order *Eumastacoidea* and to regard it as a family, without close affinity with the other families of the order. It differs from the *Chorotypidae* in such essential characters as presence of strongly sclerotized penis, dorsal position of the ejaculatory duct, clear differentiation of ecto- and endophallus, and position of endophallic sacs. All these characters of the *Eruciidae* are shared with most of the other families of the order, but the homology and even the analogy of most of the parts of the phallic complex are doubtful and need further study.

Distribution: Oriental Region, most of Austral-Asian islands (including New Guinea).

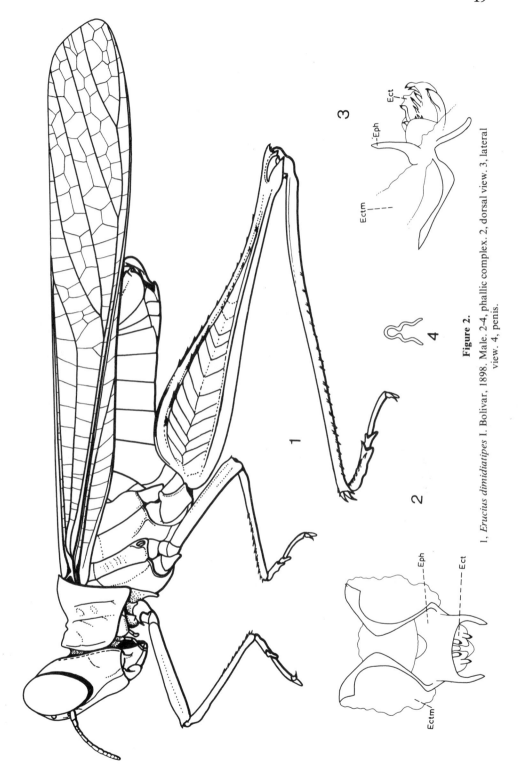

Figure 2.

1, *Eructus dimidiatipes* I. Bolivar, 1898. Male. 2–4, phallic complex. 2, dorsal view. 3, lateral view. 4, penis.

The family contains the following subfamilies:

1. *Chininae* Burr, 1903
2. *Eruciinae* Burr, 1903
3. *Mastacideinae* Rehn, 1948

Subfamily *Chininae,* was separated from *Eruciidae* on the basis of rather weak external characters, and it seems cannot be very well separated on the present family level. The same is applicable to the subfamily *Mastacideinae.*

Family
Gomphomastacidae

(Fig. 3)

Diagnosis: Body subcylindrical. Head almost vertical; face flattened; costa frontalis sulcate or subsulcate, sometimes unicarinate below ocellus. Basi-occipital slit absent; dorsal cervical membrane without sclerotized plates. Antennae short or moderately long, filiform or clavate, 9-26 segmented. Apterous or brachypterous. Hind femora not or slightly compressed; spurs of hind tibia calcariform, normally developed on both sides or second external spur strongly reduced or absent and second internal spur reduced, or second spur on both sides reduced; hind metatarsus serrated, spined or unarmed.

PHALLIC COMPLEX: Ectophallus large membraneous, capsule-like, covering whole endophallus, in distal end with pair of appendages. Endophallus clearly differentiated from ectophallus, relatively large; penis moderately large, strongly sclerotized. Endophallic sacs large; ejaculatory sac in ventro-proximal position; ejaculatory duct connected with the sac in dorso-proximal part; spermatophore sac in dorso-distal position, large and well separated from ejaculatory sac; phallotreme short and wide. Epiphallus bridge-shaped, arculate, with pair of latero-distal appendages with hooks at apices.

Spermatheca: Simple ampoule-like, pear-shaped or almost spherical.
Karyotype: $2n\,\sigma = 17, 19, 23$. Too little data exists to decide which karyotype is predominant.
Type genus: *Gomphomastax* Brunner, 1898.

The family at present is divided into the four following subfamilies:

1. *Biroellinae*
2. *Gomphomastacinae*
3. *Morseinae*
4. *Teicophryinae*

Distribution: Palaearctic, Nearctic, Oriental and Australian Regions.

The family *Gomphomastacidae* is not so sharply defined and not so clearly separated as the other families of the superfamily. It has many common and transient characters with the family *Eumastacidae* (but not with other families). The main morphological character separating them is that in *Gomphomastacidae* the epiphallus is bridge-shaped, while in *Eumastacidae* it is shield-shaped and even this division sometimes is not very clear.

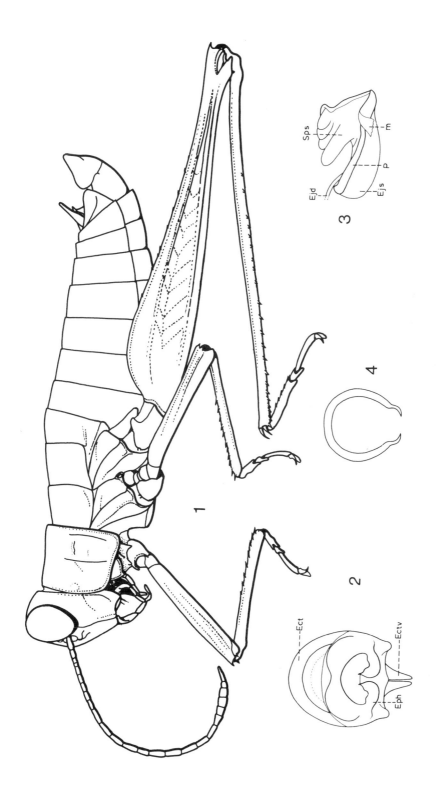

Figure 3.

1, Gomphomastax clavata (Ostroumoff, 1881). Male. 2-4, phallic complex. 2, dorsal view. 3, the same, lateral view. 4, penis.

The other feature separating these two families is that the *Gomphomastacidae* are distributed in the Eastern Hemisphere. While *Eumastacidae* are confined to the Western hemisphere.

The subfamilies of the family were created on the basis of external characters, which sometimes are of uncertain value.

Family

Morabidae

(Fig. 4)

Diagnosis: Body elongate-cylindrical, stick- or straw-like. Head horizontal or semi-horizontal; costa frontalis low, sulcate or subsulcate, with lateral carinae present or partly oblitereated. Basi-occipital slit present; dorsal cervical membrane with paired, sclerotized plates. Antennae ensiform, 8-18 segmented. Fully apterous. Hind femora not compressed; second spur on inner side of hind tibia reduced, spiniform, not articulated, or absent; second spur on outer side of hind tibia absent; hind metatarsus not toothed or serrated, but bearing small tubercle.

PHALLIC COMPLEX: Ectophallus large, membraneous, capsule-like, in apical part sclerotized and bearing tooth-like distal projections. Endophallus clearly differentiated, relatively large, with penis strongly sclerotized, forming U-shaped fork, with prongs directed towards distal end. Endophallic sacs large; ejaculatory sac in ventro-proximal position; ejaculatory duct connected with the sac in dorso-distal part; spermatophore sac in dorso-distal position, relatively small, poorly differentiated from ejaculatory sac and from phallotreme; phallotreme wide, opened between apical valves of endophallus. Epiphallus shield-shaped, in middle weakly sclerotized, with pairs of strongly sclerotized apical projections.

Spermatheca: Narrow conical, with base of conus at the proximal end.
Karyotype: $2n\sigma = 13, 15, 17, 19, 21$. Predominant number $2n\sigma = 17$.
Type genus: *Moraba* Walker, 1870.

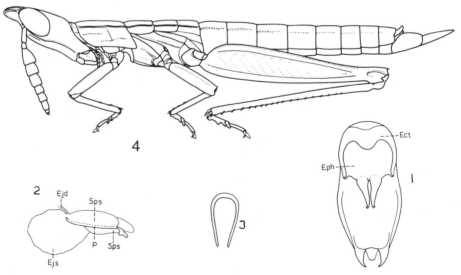

Figure 4.

4, *Moraba scurra* Rehn, 1952. 1-3, phallic complex. 1, dorsal view. 2, lateral view. 3, penis.

Distribution: Australia and Tasmania.

Morabidae differ from other families of the superfamily by their bacilliform body, ensiform antennae, reduction or loss of the second spur of hind tibia, shape and position of the head accompanied by the presence of a basi-occipital slit and sclerotized plates of dorsal cervical membrane. All these characters separate the *Morabidae* from other eumastacoids.

However, the structure of the phallic complex is rather similar to that in *Gomphomastacidae,* particularly in the genus *Gomphomastax.* The similarity indicates a more close relationship between these families than can be drawn from the rather diverse external morphological characters.

The family is comparatively uniform in its characters and is not divided into subfamilies.

Family

Eumastacidae

(Fig. 5)

Diagnosis: Body subcylindrical. Head subvertical; face moderately or strongly flattened; costa frontalis sulcate, sometimes strongly, or subsulcate. Basi-occipital slit absent; dorsal cervical membrane without sclerotized plates. Antennae short, filiform, 8-15 segmented. Macropterous, with reduced wings, or apterous. Hind femora not compressed or slightly compressed; spurs on both sides of hind tibia normally developed, calcariform; dorsal or dorso-lateral side of hind metatarsus with one or several spines.

PHALLIC COMPLEX: Ectophallus moderately large, membraneous or slightly sclerotized, capsule-like, covering whole endophallus and not specialized, at distal end with pair or several pairs of appendages. Endophallus clearly differentiated, relatively large; penis relatively large and robust, strongly sclerotized, forming U-shaped fork with prongs directed towards distal end. Endophallic sacs large; ejaculatory sac in ventro-proximal position; ejaculatory duct connected with the sac in dorso-proximal part; spermatophore sac in dorso-distal position, relatively large and well differentiated from ejaculatory sac; phallotreme large, with wide opening. Epiphallus shield-shaped, with distal appendices, sometimes, proximal or distal margins, or both, of epiphallus are incurved, giving to the shield a wide-bridged appearance.

Spermatheca: Elongate-, ampoule- or sac-shaped, sometimes with short basal bulge.
Karyotype: 2n $\circlearrowleft$ = 19, 21. Only a few data concerning the karyotype are available, to determine the predominant type.
Type genus: *Eumastax* Burr, 1903.

The following subfamilies are considered here as belonging to the family:

1. *Epistacinae*
2. *Espagnolinae*
3. *Eumastacinae*
4. *Paramastacinae*
5. *Parepistacinae*
6. *Pseudomastacinae*
7. *Temnomastacinae*

The *Espagnolinae* is placed into the family only tentatively, owing to the lack of material for studying genital organs of the subfamily.

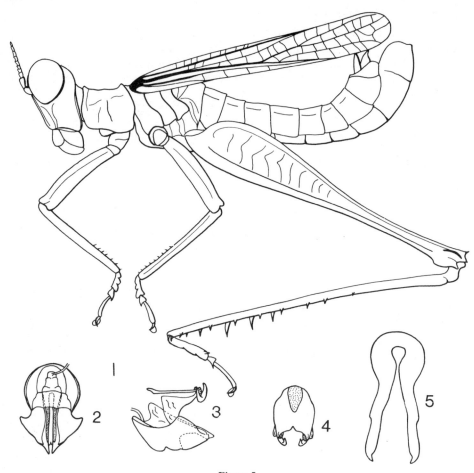

Figure 5.

1, *Eumastax surda* Burr, 1899 (after Dirsh, 1961). 2-5, phallic complex of *Eumastax bouvieri* C. Bolivar, 1918. 2, dorsal view; 3, the same, lateral view; 4, epiphallus; 5, penis, dorsal view. (After Descamps, 1971).

Distribution: Neotropical and South Nearctic Regions.

The characters of the subfamilies are rather uncertain and probably further study and more material will lead to the synonymy of most of them.

This family, representing the largest assemblage of subfamilies, is very clearly separated from all other families of the order *Eumastacoidea,* except the family *Gomphomastacidae* with which it has a certain amount of common characters. However, the structure of the epiphallus and shape of the penis divide them sufficiently to consider them as the separate families.

Family

Euschmidtiidae

(Fig. 6)

Diagnosis: Body subcylindrical. Head subvertical; face flattened; costa frontalis sulcate or subsulcate. Basi-occipital slit absent; dorsal cervical membrane without sclerotized plates. Antennae short, filiform, 15-19 segmented. Macropterous, with reduced wings or apterous. Hind femora not compressed or slightly compressed; external spurs of hind tibia developed, normally, calcariform; second internal spur much reduced sometimes spiniform or absent; hind metatarsus with one or several dorso-lateral spines.

PHALLIC COMPLEX: Ectophallus relatively large membraneous or weakly sclerotized, capsule-like, of simple structure, not specialized. Endophallus clearly differentiated, relatively large and highly specialized; penis large, strongly sclerotized, of complicated structure, monoscleritic and fused at distal end, or divided into two sclerites. Ejaculatory sac very small in dorso-distal position; ejaculatory duct connected with the sac in dorsal part; spermatophore sac very large, in ventro-proximal position, well differentiated from ejaculatory sac; phallotreme short and narrow. Epiphallus represented by elongated disc with lateral, strongly sclerotized marginal bars, at distal ends with pair strongly sclerotized hooks.

Spermatheca: Simple ampoule-like, pear-shaped.
Karyotype: $2n\sigma = 19, 21, 25$. Predominantly with $2n\sigma = 21$.
Type genus: *Euschmidtia* Karsch, 1889.

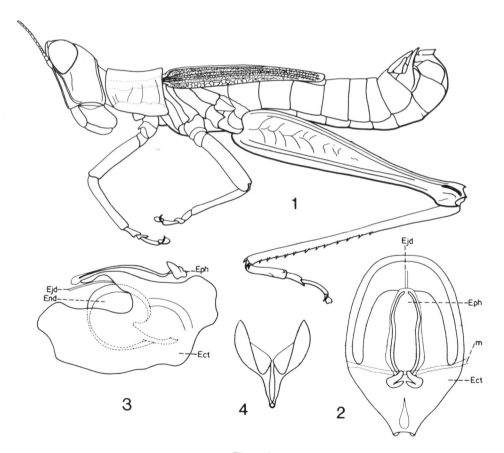

Figure 6.

1, *Euschmidtia burri* Uvarov, 1953. Male. 2-4, phallic complex. 2, dorsal view. 3, lateral view. 4, penis.

The family contains the following subfamilies:

1. *Euschmidtiinae* Rehn, 1948
2. *Malagassinae* Rehn & Rehn, 1945
3. *Miraculinae* I. Bolivar, 1903
4. *Pseudoschmidtiinae* Descamps, 1964

The subfamilies *Miraculinae* and *Malagassinae* were rightly synonymised by Descamps (1964). The remaining three subfamilies are rather distinct. *Euschmidtiinae* differs from *Pseudoschmidtiinae* by the shape of the penis with its prongs connected at distal end, while in the *Pseudoschmidtiinae* they are separated. The *Miraculinae* differs from both of the subfamilies mainly by their extraordinary appearance.

Distribution: Ethiopian Region, Madagascar and Comoro Islands.

In external morphological characters the *Euschmidtiidae* range from very simple, ordinary appearance, similar to many eumastacids, to an extravagantly armoured and sculptured body. The structure of the phallic complex, is highly distinct, particularly in the structure of the endophallus, with large and complicated shape of penis and relatively small ejaculatory sac in comparison with very large spermatophore sac. The whole phallic complex differs from that in other families of *Eumastacoidea* that it is difficult at the present to find a relationship between them. As a very remote possibility the affinity with the family *Eumastacidae* may be suggested.

Family

Chorotypidae

(Fig. 7)

Diagnosis: Body compressed. Head subvertical; face flattened; costa frontalis low, or partly obliterated, sulcate or subsulcate on whole length. Basi-occipital slit absent; dorsal cervical membrane without sclerotized plates. Antenna filiform, short, 12-14 segmented. Tegmina and wings mostly fully developed, sometimes reduced or absent. Hind femora compressed. Both spurs on both sides of hind tibia calcariform; dorsal margins of hind metatarsus toothed on whole length.

PHALLIC COMPLEX: Ectophallus large, forming strongly sclerotized, and highly specialized capsule, in apical part capsule forming large teeth, lobes and projections; endophallus not clearly differentiated and not sclerotized; penis absent or poorly developed. Endophallic sacs large, ejaculatory sac in ventro-distal position, ejaculatory duct connected with the sac in ventro-proximal part; spermatophore sac in dorso-proximal position; phallotreme short and wide, opened between "lophi of epiphallus". Epiphallus (?) strongly sclerotized, at apex forming mushroom-like structure, middle part forming long, narrow, stalk connected with two-branched basal part (lophi?), each branch ending in robust, upcurved hook with numerous teeth at apex.

Spermatheca: Ampoule-like, pear-shaped.
Karyotype: $2n\,\sigma = 21$.
Type genus: *Chorotypus* Serville, 1839.

Distribution: Ethiopian and Oriental Regions.

Chorotypidae are so different from the other families of *Eumastacoidea,* that as early as 1876 Stal considered them as a family. Burr (1903) lowered it to subfamily rank and has remained so up to the present time.

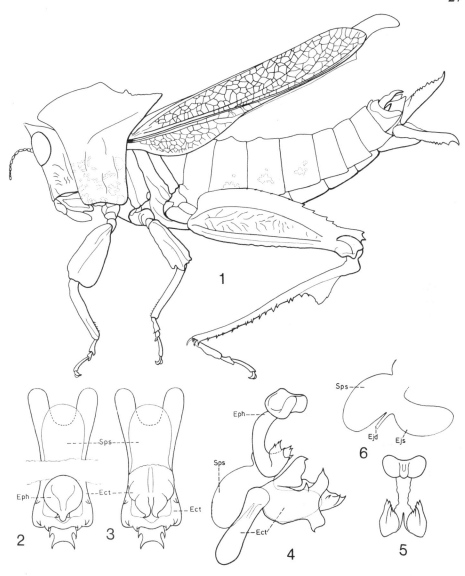

Figure 7.

1, *Hemicharianthus batesi* Rehn & Rehn, 1945. 2-6, phallic complex of *Erianthus guttatus* (Westwood, 1841). 2, dorsal view. 3, the same (membrane and epiphallus removed). 4, the same, lateral view. 5, epiphallus. 6, endophallic sac.

Erianthinae Burr (1903) were synonymised with *Chorotypinae* by C. Bolivar in 1932 on the basis of the external characters of both subfamilies. The study of the phallic complex (Dirsh, 1965) confirmed this synonymy.

Phallic complex of the family is so different from the other families of the order, that homology and even analogy of the parts can be considered only as a tentative arrangement adopted for the sake of description of the morphological structures. The true morphological and functional meaning of some parts can only be rightly or wrongly guessed, and needs to be studied on fresh material and checked with live specimens.

The main differences of the family from other eumastacids are the absence of a clear differentiation between the ecto- and endophallus, absence or presence of a primitive sclerotized penis, the ventral position of the ejaculatory duct opening, and upper position of the phallotreme opening between the "lophi of epiphallus". The shape and position of the epiphallus cast doubts if it is a structure analogous to the epiphalli of other eumastacids.

Probably, the family, if it has any affinity with other eumastacids, it is a very remote one. It is possible to speculate that it branched from the Pro-Orthopteroidea stock independently from the other eumastacids.

Superfamily

Proscopioidea

(Fig. 8)

Diagnosis: Body elongated, stick-like, mostly cylindrical. Head narrow-conical mostly elongated, horizontal; face incurved; costa frontalis only traceable or subsulcate. Basi-occipital slit present; dorsal cervical membrane without sclerotized plates. Antennae short, filiform. Antennal organ present, mostly on apical segment. Protorax tube-like. Prosternal process absent. Tegmina and wings absent or strongly reduced. Hind legs weakly saltatorial almost cursorial. Brunner's organ absent or rudimentary. Second spur on both sides of hind tibia strongly reduced. Dorsal side of metatarsus sulcate, sometimes with serrated margins.

Sound producing mechanism not found, but they produce (in cages) high-pitched squeaks. The method of producing this sound is unknown, (Mason, 1969).

PHALLIC COMPLEX: Ectophallus membraneous or weakly sclerotized, capsule-like, not specialized; endophallus clearly differentiated. Penis present, monosclerited, rod-like. Endophallic sacs small. Ejaculatory sac very small; ejaculatory duct connected with ejaculatory sac in dorso-proximal part; spermatophore sac elongated, tubular, continuing into narrow phallotreme. Epiphallus strongly sclerotized, bridge-shaped, with hook-like, incurved distal ends.

Spermatheca: Elongate, sac-like, with two basal diverticula and short, as wide as the spermatheca, spermatecal duct.
Karyotype: 2n σ^7 = 17, 19.
Type genus: *Proscopia* Klug, 1820.

Contains a single family *Proscopiidae,* which is not subdivided into subfamilies.

Distribution: South America.

Proscopiidae recently were considered as a family of *Acridoidea.* Dirsh (1966) transferred it into superfamily *Eumastacoidea.* Sharov (1968) for unknown reason placed it into superfamily *Pneumoroidea.*

In this work *Proscopiidae* are considered as a superfamily of the order *Eumastacoidea.* The reason for placing them into this order are the following common characters they share with the eumastacids: absence of tympanal organ (as primary character), presence of the antennal organ, absence of prosternal process, monoscleritic penis, position of the opening of ejaculatory duct and highly specialized endophallus.

Proscopioidea is considered as an equal superfamily but not a family of the *Eumastacoidea* on the reason of the following characters of difference: they are not

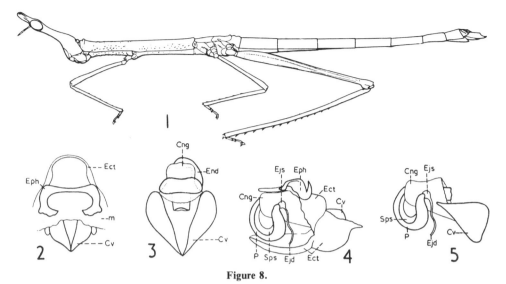

Figure 8.

1, *Proscopia scabra* Klug, 1820. Female. 2-5, phallic complex. 2, dorsal view. 3, the same but membrane and epiphallus removed. 4, lateral view, the same as fig. 2. 5, endophallus.

possessing ileal coeca, they possess of unique and different from *Eumastacoidea* structure of prothorax; in details of the structure of the phallic complex and in spermatheca they also show a marked difference.

Blackith and Blackith (1968) compared 92 characters of the families *Eumastacidae* and *Proscopiidae*, and found that 49 characters are shared by both families and 43 are not shared.

Superfamily

Trigonopterygoidea

(Fig. 9)

Diagnosis: Body compressed. Head acutely conical, vertical; face incurved; clear fastigial furrow present, fastigial foveolae or areolae absent, costa frontalis subsulcate. Dorso-occipital slit absent; dorsal cervical membrane without sclerotized plates. Antennae moderately long, ensiform. Antennal organ rudimentary or absent. Pronotum rectangular in cross section. Prosternal process present. Tympanum absent. Tegmina and wings mostly fully developed (in one genus shortened). Hind femur slightly compressed, with lower basal lobe shorter than upper one. Brunner's organ present. Hind tarsi dorsad convex, unarmed.

PHALLIC COMPLEX: Whole complex is in reverse position, dorsal side turned ventrad, with penis directed towards anterior end of body and epiphallus in ventral position. Ectophallus simple, capsule-like, membraneous or with weak sclerotizations, not specialized. Endophallus highly differentiated and specialized, of complicated structure. Penis bisegmented, with apical valves bilobate and fused in basal part. Ejaculatory sac small, in dorso-proximal position; ejaculatory duct connected with ejaculatory sac in dorso-proximal part (it is analogous to ventro-proximal in *Eumastacoidea*). Spermatophore sac relatively large; phallotreme large and wide. Epiphallus transverse, disc-shaped, well sclerotized, with numerous teeth.

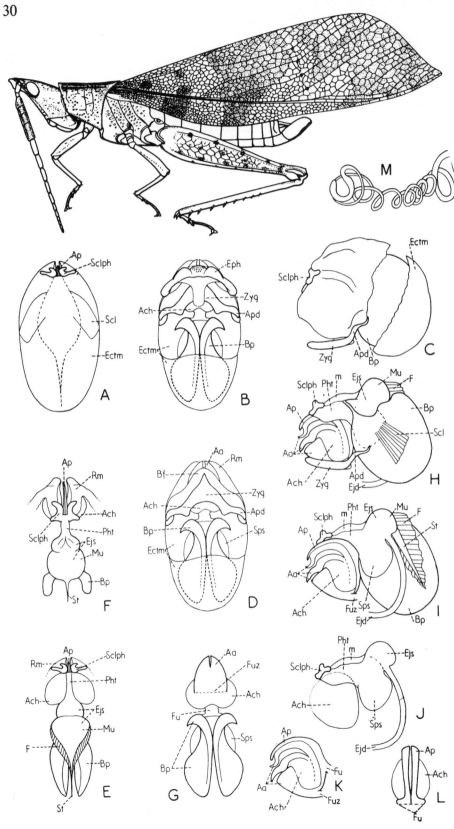

Spermatheca with apical reservoir of irregular form; spermathecal duct exceptionally long, gradually widening towards basal end.

Karyotype: 2n ♂ = 23. (Based on insufficient amount of data).

Type genus: *Trigonopteryx* Charpentier, 1841.

Contains a single family *Trigonopterygidae* which is not subdivided into subfamilies.

Distribution: South of Oriental Region and Austral-Asian Islands.

Trigonopterygoidea are entirely isolated from the other superfamilies of the order. However tentatively they are placed here into *Eumastacoidea* on the ground of several common characters: The primary absence of tympanal organ, widening towards apex tegmina, wide remigium of hind wings, and position of the ejaculatory sac and duct.

The characters of difference include presence of the prosternal process, reversed phallic complex structure of which have not analogy in the other superfamilies of the order, and in the order *Acridomorphoidea* either. The homology and analogy of the most parts of the phallic complex is only supposition for sake of necessity to name them at least tentatively. Function of some parts, such as the ejaculatory duct, ejaculatory and spermatophore sacs is undoubtedly the same as in *Eumastacoidea* and *Acridomorphoidea*, but in other complicated arranged structures, the function may be only guessed or remains unknown.

The description of the whole phallic complex of *Systella rafflesii* Westwood, 1841 is given (Dirsh, 1956) for explanation parts of the structure.

Westwood (1841) describing genus *Systella* placed it "between *Truxalides* and *Conophori*". Walker (1870) considered the two known genera — *Trigonopteryx* and *Systella* as a separate family and named it *Trigonopterygidae*. I. Bolivar (1884) lowered the rank of the group and placed it into *Pyrgomorphinae* as a subtribe. In 1909 I. Bolivar was still regarding them as a member of *Pyrgomorphinae*, but as a "Sectio *Systellae*." Dirsh (1952) reinstalled the group as a separate subfamily *Trigonopteryginae* following Walker's definition and in 1966 restored them as a Walker's family.

Kevan (1952) divided *Trigonopterygidae* into three tribes — *Trigonopterygini*, *Borneacridini* and *Xyronotini*, with single genus *Xyronotus*. The latter he considered as "temporarily attached to the *Trigonopteryginae*". Dirsh (1955) removed *Xyronotus* from the *Trigonopteryginae* and installed it first as a separate, unrelated, subfamily and later (Dirsh, 1956) raised it to family rank.

In 1966 Dirsh raised it into a superfamily rank equivalent to the three superfamilies of the *Acridomorpha*.

In present work the superfamily is removed from the order *Acridomorphoidea* and placed into order *Eumastacoidea*, in the same superfamily rank.

In the phylogenetic scheme Dirsh (1956) considered origin of *Trigonopterygoidea* from the early *Eumastacoid* stock on the basis of the whole complex of known characters. Helwig (1958) on the basis of the karyotypes of *Acridoidea* came in this respect to the same conclusion.

◀ **Figure 9.**

Trigonopteryx hopei (Westwood, 1841). Male. A-L. Phallic complex of *Systella rafflesii* (Westwood, 1841). A, B, whole complex, dorsal and ventral views; C, D, phallic organ, lateral and dorsal views, epiphallus removed; E, F, phallic organ, dorsal view, epiphallus, ectophallic membrane, zygoma and apodemes removed (in F, distal end raised to show apical valves of penis and sclerotised parts of phallotreme); G, as for E, but ventral view; H. phallic organ, lateral view, ectophallic membrane removed; I, phallic organ lateral view, in longitudinal section; J, diagram of endophallic sac; K, phallic chamber; L, apical part of penis, dorso-apical view; M, spermatheca.

ORDER

Acridomorphoidea

Diagnosis: Body of various shape. Distal antennal organ absent or present. Tympanal organ present or secondary absent. Tegmina and wings fully developed, reduced or absent. Brunner's organ present. Dorsal sides of hind leg metatarsus not armed.

PHALLIC COMPLEX: From highly specialized to relatively simplified. Penis bisclerited.

Main characters differentiating *Acridomorphoidea* from *Eumastacoidea* are: primary presence of tympanum (which sometimes is lost as secondary development). Unarmed dorsal side of hind leg metatarsus; and bisclerited penis.

Order *Acridomorphoidea* contains the following superfamilies:

1. *Acridoidea*
2. *Pamphagoidea*
3. *Pneumoroidea*

The *Acridomorphoidea* probably originated from the hypothetical *Protorthoptera* stock, from which they have branched probably later than *Eumastacoidea*, possibly in the Permian period, when the *Eumastacoidea* were already flourished. This point of view of course cannot be proved on factual material, since fossil remnants of *Acridomorphoidea* almost not existing.

Key to superfamilies

1 (2) Ectophallus membraneous or partly weakly sclerotized, not forming cingulum. Endophallus not differentiated, not sclerotized or partly slightly sclerotized. Epiphallus weakly sclerotized, disc-shaped, without ancorae and lophi; oval sclerites absent, sound-producing mechanism of femora-abdominal type.

Pneumoroidea

2 (1) Ectophallus partly or almost fully sclerotized, capsule-like, shield-like or forming cingulum. Endophallus forming sclerotized penis. Epiphallus of various shapes, mostly strongly sclerotized; oval sclerites present or absent. Sound-producing mechanism of various types.

3 (4) Lower basal lobe of hind femur mostly longer than upper, or both of the same length. Spermatophore sac in dorsal or ventral position. Fastigial furrow present, gonopore processes absent. Oval sclerites mostly absent.

Pamphagoidea

4 (3) Lower basal lobe of hind femur mostly shorter than upper or both almost of the same length. Spermatophore sac mostly in middle (sometimes in dorsal) position. Fastigial furrow absent. Gonopore processes mostly present. Oval sclerites present.

Acridoidea

Superfamily

Pneumoroidea

Diagnosis: PHALLIC COMPLEX: Ectophallus membraneous, partly weakly sclerotized, not forming cingulum. Endophallus membraneous, partly weakly sclerotized, not forming clearly defined penis. Epiphallus disc-shaped, without ancorae and lophi. Oval sclerites absent.

Tegmina and wings fully developed, shortened, lobiform or absent. Tympanum absent. Sound-producing mechanism of abdomino-femoral type. Antennae short or strongly elongated.

This superfamily is divided into three families:

1. *Pneumoridae*
2. *Tanaoceridae*
3. *Xyronotidae*

Key to families

1 (2) Tegmina and wings fully developed, shortened, lobiform or absent. Antennae short.

Pneumoridae

2 (1) Completely wingless. Antennae short or elongated, longer than body.

3 (4) Body compressed. Antennae short. Male cercus trifurcate.

Xyronotidae

4 (3) Body subcylindrical. Antennae longer than body. Male cercus simple, subconical.

Tanaoceridae

The phallic complex of *Pneumoroidea* is very different from that of *Acridoidea* studied to date, and the superfamily may be distinguished by this character alone. But there are several characters of similarity: the presence of the epiphallus and similar differentiation into ectophallus and endophallus. These characters connect Pneumorids with the rest of *Acridoidea* in their general features. It is interesting to note, however, that the *Pneumoroidea* phallic complex is remotely similar to the corresponding structure in *Tettigonioidea* and *Gryllacridoidea* (Snodgrass, 1937). If one considers that *Tettigonioidea* and *Gryllacridoidea* are more primitive than *Acridoidea*, then it may be inferred that the *Pneumoroidea* are more primitive than the other *Acridoidea*.

Pneumoroidea display a very high degree of specialization in their sound-producing mechanism. This specialization in the *Pneumoroidea* is probably of very ancient origin.

When and how this sound-producing mechanism originated in the *Pneumorids* can be solved only on the basis of fossil material, which at present is lacking. It should be noted that analogous stridulatory mechanisms exist in certain groups of *Gryllacridoidea*, but this could be coincidental and an independent parallel development. A second possibility is that this character was primarily developed in a common ancestral stock of Orthopteroid insects and was retained in *Pneumoridae, Tanaoceridae* and *Xyronotidae,* reaching the highest point of specialization in the *Pneumoridae*.

Another interesting point is that although *Pneumoroidea* produce tremendous noise they do not possess a tympanal organ, which is considered as an organ of sound perception. Probably they have some other kind of organ for sound perception, as yet unknown. It is known, however, that there is a great variety of these organs in various groups of insects (Haskell, 1961). The tympanal organ exists only in Orthopteroids, Hemiptera and Lepidoptera. In *Acridoidea* it is present in six out of fourteen families. It is absent in all the families with an abdomic-femoral sound-producing mechanism.

34

In *Gryllacridoidea*, the group possessing the abdomino-femoral mechanism, the tympanal organ (on the front tibia) is absent, while in some groups, without the mechanism, the tympanal organ is present. As the tympanal organ is present in *Acridoidea* in the more advanced families and subfamilies, one can conclude that the groups that lack it are more primitive.

Family

Pneumoridae

(Fig. 10)

Diagnosis: Body compressed or cylindrical, abdomen in males mostly grossly inflated. Head short, above subglobular; face, in profile, straight, vertical or inclined; frontal ridge absent; fastigium of vertex short, at apex widely rounded; fastigial foveolae or areolae absent; fine fastigial furrow present. Antennae short, filiform. Dorsum of pronotum crest-shaped or tectiform, crossed by four sulci. Prosternal process absent. Meso- and metasternum with deep furcal sutures and very deep foveolae (sternal apophyseal pits). The mesosternal lobes (sternelli) are relatively small. Tympanum absent. Tegmina and wings fully developed, shortened or vestigial, hidden under pronotum. Venation is of primitive type. Tegmina, in macropterous males, wide.

Venation: Costa first main vein, well defined from the basal articulation of tegmen. It is located posteriorly to anterior or costal margin and reaches about half the length of tegmen.

Subcosta is very well defined from the basal articulation. It runs almost to apex of tegmen and is unbranched.

Radius, third main vein which is very well defined from the basal articulation; and forms branch, Radial sector which itself forms three or four branches.

Media, derived from the basal articulation. It is two- or sometimes three-branched in the apical half.

Cubitus. It emerges from basal articulation and near base is branched into cubitus one and postcubitus or cubitus two, cubitus one is unbranched or two-branched in apical half.

Dividing vein, vena dividens, or first anal vein. The next vein after vena dividens is first vannal vein or the second anal vein.

Hind wing in macropterous males is remarkable for its large remigium, of almost the same size as vannus. Vannal fold present. Costa forms margin of wing, and Subcosta almost reaches apex. Radius and Media are fused in basal part; in apical half they are both branched, Radius into Radial sectors and Media into Media anterior and Media posterior. Cubitus one and Cubitus two are unbranched. Vena dividens or first anal vein is well pronounced and wing is flexed along it; all veins posterior to it are vannal veins (Snodgrass, 1935) or anal veins (Ragge, 1955).

In brachypterous females venation of tegmen and wing is essentially the same as in males, but first cubital vein in all female tegmina is unbranched. All veins are less strongly developed than in males and show definite signs of reduction.

In micropterous females tegmen is thickened and strongly sclerotized and, in a few, greatly reduced, but main veins can be traced. The wing in these females is completely hidden under pronotum. It is much larger than tegmen, and its venation is quite detectable, consisting of all veins as in males, but reduced and unbranched. Wing is folded singly only along vannal flex. Its reticulation is rather strong and is possibly part of sound-producing mechanism.

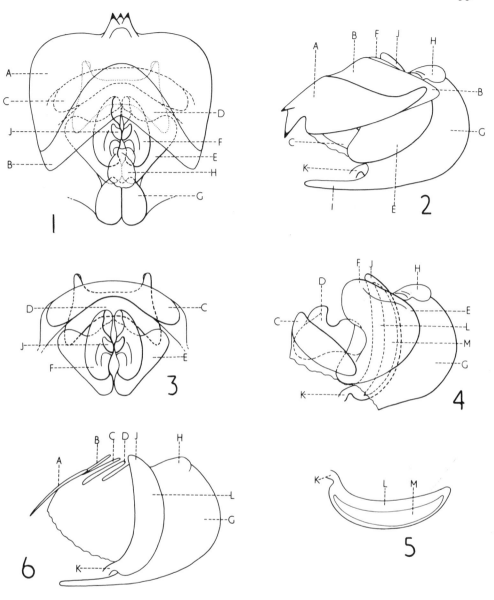

Figure 10.

Bullacris unicolor (Linnaeus, 1758), phallic complex. 1, dorsal view. 2, the same, lateral view. 3, the same, but epiphallus removed. 4, the same as fig. 3, lateral view. 5, endophallus, lateral view. 6, schematic section of phallic complex. (For explanation of parts, see text).

Net-like reticulation of tegmina and wings (archedictyon) exists in both sexes, but is obscure in the sclerotized tegmina of micropterous females. It is well pronounced, however, in the micropterous type of wings.

Front and middle pair of legs have no unusual specialisation, except that femora are sometimes tuberculate. Hind legs, differ that functionally they have lost or are losing saltatorial ability and approximate to cursorial. They are weak, short and rather slender. Lower basal lobe of hind femur longer than upper one. In the middle of inner side of male femur there is short, strongly sclerotized, longitudinal carina with series of strong, short, transverse ridges for stridulatory purposes.

Brunner's organ in most cases absent, but sometimes it can be traced as a vestigial formation, and in other cases, as in *Pneumora* and *Parabullacris*, it is fairly well developed. External apical spine of hind tibia present. Male cerci, supra-anal and subgenital plate not specialized, but paraprots are large and often exceed length of supra-anal plate.

Sound-producing mechanism in males consists of strongly sclerotized ridges on third abdominal tergite, ridges form crescent-like row, and are smaller at ends, becoming gradually larger in median part. At upper end of row there is small, tubercle-like inflation of body wall of unknown function. Lower part of row gradually diminishes to point of obliteration. In *Physophorina* and *Pneumora* the row is represented in the upper half by large, rough ridges and in lower half, with a small gap between them, by small, slender and more densely placed ridges.

Second part of stridulatory mechanism is short high carina, bearing small row of strong, transverse ridges on inner side of hind femur. Sound is produced by rubbing ridges of the abdomen with ridges of hind femur. It is assumed that the inflation of the male abdomen represents a further specialization for sound production, its function being a resonator amplifying sounds.

It is not known by what mechanism sounds are produced by females. There is possibility that they are produced by rubbing folded vannus of wing against the abdominal wall. In *Penumora inanis* posterior part of vannus of wing is covered with net of rather strong veinlets between main vannal veins, and both these main veins and veinlets in this region are covered with small, rather strong teeth. Purpose of these teeth could be stridulatory.

PHALLIC COMPLEX: Relatively small, membraneous partly weakly sclerotized. Epiphallus (A) is large, discoidal, without ancorae, lophi or oval sclerites, but with three tooth-like median projections at the anterior end or with group of small spines in middle. Lateral plates large posterior projections well developed; posterior part of the epiphallus is weakly sclerotized membrane (B), which is considered as part of epiphallus. The major part of epiphallus is connected by membraneous fold with ectophallic membrane, which forms sclerotized transverse part (C); distal end of this part has folding membraneous continuation, which is connected by fold at distal end, with weakly sclerotized disc-like part of membrane (D). Distal part of discoidal sclerotization forms membraneous fold and is connected with pair of dorsal slightly sclerotized lateral valves (E); lateral part of these valves is inflated and dorsal part protrudes upwards and forms pair of lobes (F). E valves at distal end are joined with thin-walled sac (H) of undefined form. This sac is continuation of pair of ventro-posterior valves (G). Proximal end of these valves produces plate-like membraneous continuation (I). Endophallus is represented by banana-shaped membraneous tube (L), on sides of which there is pair of longitudinal, rod-like, weak sclerotizations (M); at apex its edges merge with edges of ectophallic sac, and form pair of laterally protruding small lobes (J) between which is distal opening of endophallus. Ejaculatory duct (K) is rather wide.

The structure of the endophallus is extremely simple, without definite division into ejaculatory and spermatophore sacs.

The study of the phallic complex of adult pneumorids suggests that their endophallus could be interpreted rather as a simple widening of the vasa deferentia. The ectophallus then represents a secondary external invagination of the endophallus which has acquired a certain degree of differentiation. The epiphallus may be a derivative of this invagination. However, the possibility that it may be derived from the tergal metamere is not excluded.

Spermatheca: Divided into two groups. In the first group it is a narrow vermicular tube, with several vermicular diverticula. In second group it is a sac-like formation with several large, pocket-like diverticula.

Karyotype: 2n ♂ = 23. This is the only one case mentioned in literature.

Type Genus: *Pneumora* Thunberg, 1775.

At present only nine genera of this family are known. Here they are divided into two subfamilies.

Distribution: South, South West and East Africa.

This family is original and highly distinguishable that since Thunberg (1810) called them *Pneumorae* it retains this family name.

Family *Pneumoridae* is divided here into two subfamilies:

  1. *Pneumorinae*
  2. *Parabullacrinae*

There is no fossil evidence which can help to establish the relationship of the pneumorids with the other families of *Acridoidea* or with the other groups of Orthoptera. At present, only the study of the comparative morphology of the group can provide some indirect clues. Rehn, 1941 expressed the opinion that the Pneumorids are an ancient group equivalent to the *Tetrigoidea*. Smart, 1953 stated that the wing venation of Pneumorid males is "remarkably primitive", implying primitiveness of the family. Ragge, 1955 considered the wing venation of Pneumorids as the most primitive of all *Acridoidea*. He considered them as derived from the general *Acridoidea* stock even earlier than *Locustopsidae*, and earlier than the other branches of *Acridoidea*.

There are several characters which separate pneumorids from *Acridoidea*, making them a distinctly isolated group.

The wing venation as already explained is extremely simply in the pneumorids. Ragge, 1955 showed that morphologically they are extremely close to the fossil *Palaeodictyoptera, Stenodictyolobata* (Fam. *Dictyoneuridae*). But it is not conceivable that they were derived directly and primarily from the *Palaeodictyoptera* stock, as they have too many other characters in common with *Acridoidea* in the recent concept of this superfamily.

Thus, on one hand the venation of the pneumorids is very similar to that of the *Dictyoneuridae* and on the other hand it is clearly related to that of *Acridoidea*. This can lead to only two possibilities: firstly, that the pneumorids retained the characters of the ancient *Palaeodictyoptera* during their phylogeny and branched independently from the *Acridoidea*, but developed a parallel set of characters; and secondly, that the wings of pneumorids are the result of secondary simplification. In the latter case the pneumorids may have branched early from the common *Acridoidea* stock and possibly lost or reduced the function of the wings; as a result the wings would degenerate to a simplified form, very similar to the primitive *Palaeodictyoptera Dictyoneuridae*.

The main differences between the tegmina and wings of *Pneumoridae* and those of the *Acridoidea* are, in macropterous species, as follows:

Pneumoridae
Tegmen: Relatively very wide.

Acridoidea
Tegmen: Relatively narrow.

Vannal fold absent.	Vannal fold present.
Main veins, in apical half, curved towards posterior margin.	Main veins comparatively straight and very little or not at all curved towards posterior margin.
Remigium relatively large.	Remigium relatively small.
Vannus very small and narrow.	Vannus relatively larger and much wider.
Radial vein and radial sector in basal two-thirds fused.	Radial vein and radial sector very close together in basal half, but not fused.
Intercalary veins absent.	Intercalary veins mostly present.
Wing: Remigium relatively very large, almost as large as vannus.	*Wing:* Remigium relatively small, much smaller than vannus.
Vannus relatively very small.	Vannus large.

Subfamily

Pneumorinae

(Fig. 11)

Diagnosis: Body cylindrical or subcylindrical, in males greatly inflated, large size. Head short; face in profile, straight, slightly inclined; fastigium of vertex short, at apex rounded; Tegmina and wings in males fully developed, in females shortened or strongly shortened, hidden under pronotum. Hind femora strongly shortened, slender.

Phallic complex and spermatheca as in family description.

Karyotype: $2n\,\vec{O} = 23$.
Type genus: *Pneumora* Thunberg, 1775.

Distribution: See distribution of the family.

Subfamily

Parabullacrinae

(Fig. 12)

Diagnosis: Body compressed laterally, medium size, not inflated in males. Head short; face, in profile, slightly excurved and inclined; fastigium of vertex short, at apex rounded; Tegmina and wings in both sexes, vestigial, hidden under pronotum. Hind femora slender, relatively long, particularly in males.

Vestigial sound-producing mechanism, of the same type as in *Pneumorinae*.

Phallic complex and spermatheca, see family description.

Karyotype: Unknown.
Type genus: *Parabullacris* Dirsh, 1963.

Distribution: South and South West Africa.

Three genera of this subfamily are known all of them described by Dirsh, 1963. The same author in 1965 expressed the opinion that they may be a neotenic form of *Pneumoridae*, however, up to date there is not any positive evidence for this possibility.

This subfamily is closely related to the subfamily *Pneumorinae*. The vestigal traces of the same type of stridulatory mechanism suggests that they probably derived from the latter subfamily.

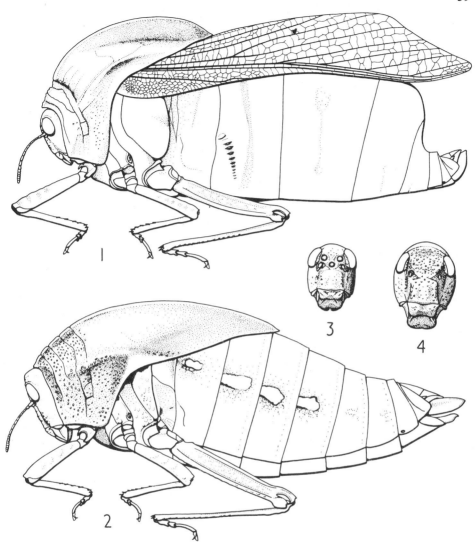

Figure 11.

Bullacris unicolor (Linnaeus, 1758). 1, male. 2, female. 3, male face. 4, female face.

Family

Xyronotidae

(Fig. 13)

Diagnosis: Body of medium size, compressed laterally. Integument rugose. Head acutely conical; face, in profile, incurved; fastigium of vertex protruding forwards, angular; fastigial furrow present; fastigial foveolae or areolae absent. Antennae slightly ensiform. Dorsum of pronotum tectiform, almost crested. Prosternal process present. Mesosternal interspace short, open. Tegmina, wings and tympanum absent. Hind femora slender, with lower basal lobe slightly shorter than upper. Brunner's

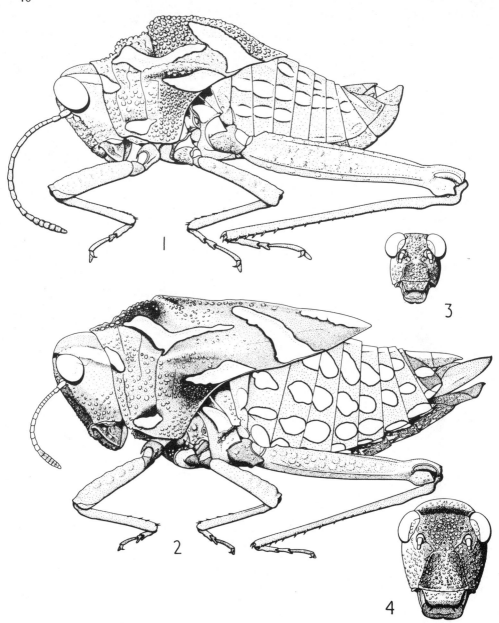

Figure 12.

Pneumoracris browni Dirsh, 1963. 1, male. 2, female. 3, male face. 4, female face.

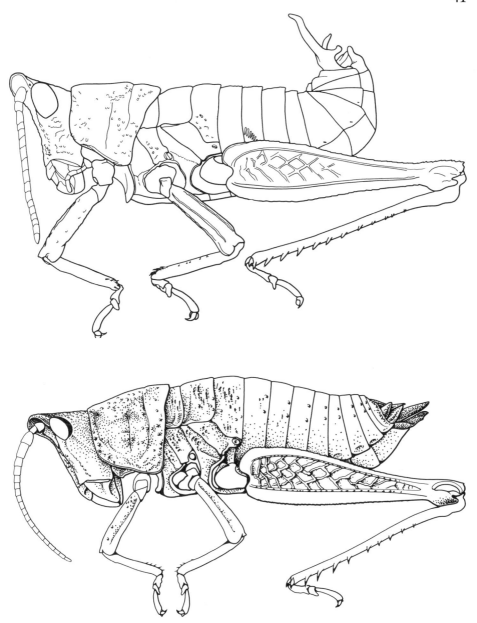

Figure 13.

Xyronotus aztecus I. Bolivar, 1884. Above male, below female.

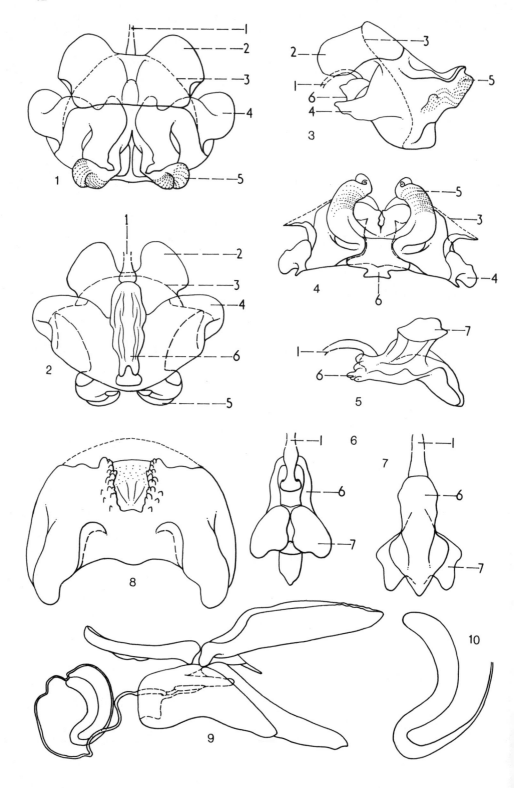

organ present. External apical spine of hind tibia present. Male cercus trifurcate; supra-anal plate angular; subgenital plate at apex bifurcate. Ovipositor moderately small, valves straight, slender.

Sound-producing mechanism represented by row small, transverse ridges on sides of third abdominal tergite; second part of the mechanism is serrated ridge on inner side of hind femora.

PHALLIC COMPLEX: Ectophallus membraneous, except slightly sclerotized dorso-proximal paired plates, which is extended and connected with pair of distal, slightly sclerotized valves; endophallus represented by undifferentiated endophallic sac and pair endophallic dorsal sclerotizations. Epiphallus shield-shaped; proximal part in middle membraneous, on both sides of the membrane pair of horn-like, more strongly sclerotized, convexities covered with ridges and tubercles are protruding, they can be considered as structures analogous to ancorae; distal part bears pair large valves analogous to lophi. Oval sclerites absent.

Spermatheca: Simple, consisting narrow, curved main reservoir and simple narrow spermathecal duct. No diverticula present.
Karyotype: $2n \sigma = 23.$
Type genus: *Xyronotus*. I. Bolivar, 1884.

The family represented by a single type genus. Only two species of the genus are known — *X. aztecus* I. Bolivar, 1884 and second species which is not named and is under description by Prof. I. J. Cantrall of Michigan University.

Distribution: Found in Mexico only.

I. **Bolivar** (1909) placed *Xyronotus* into "Sectio" *Xyronoti* of the subfamily *Pyrgomorphinae*. Kevan (1952) regarded it as a tribe which he temporarily attached to *Trigonopterygidae*. It was raised to family rank by Dirsh, 1955.

Owing to the similarity in structure of the phallic complex and a common type of the sound-producing mechanism, *Xyronotidae* together with two other families were placed into superfamily *Pneumoroidea*. The interrelation of *Xyronotidae* with other families of the superfamily is not clear. The only suggestion can be made that the family is one of the ancient relics of the ancestral stock, which miraculously survived up to the present time.

Family

Tanaoceridae

(Fig. 15)

Diagnosis: Body small, subcylindrical. Integument slightly rugose. Head subglobular; face, in profile almost straight, slightly inclined backwards; fastigial furrow and foveolae absent. Antennae filiform, very long (in male longer than body).

◄

Figure 14.

Xyronotus aztecus I. Bolivar, 1884. 1-8, phallic complex. 1, from above; 2, from below; 3, lateral view; 4, posterior view; 5, endophallus, lateral view; 6, the same, dorsal view; 7, the same, ventral view; 8, epiphallus, dorsal view; 9, skeletal parts of ovipositor and spermatheca; 10, reservoir of spermateca.

(Parts of the phallic complex: 1, ejaculatory duct. 2, dorso-proximal sclerotization. 3, dorsal membraneous part of ectophallus. 4, ventro-lateral sclerotization. 5, apical valves of ectophallus. 6, endophallic sac. 7, endophallic dorsal sclerotization).

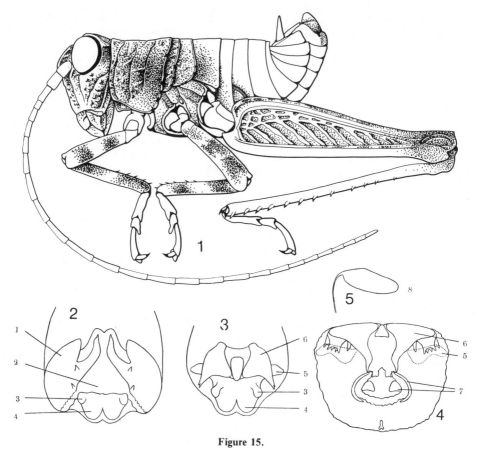

Figure 15.

Tanaocerus koebelei Bruner, 1906. 1, male. 2, phallic complex, dorsal view. 3, the same, but epiphallus removed. 4, posterior view. 5, spermatheca.

1, epiphallus. 2, membraneous part of ectophallus. 3, lateral convexities of ectophallic valves. 4, ectophallic valves. 5, lateral extension of ectophallic valves. 6, ectophallic dorsal sclerotization. 7, endophallic distal sclerotization. 8, spermatheca. (Figs. 2-5 after Grant, 1960. Modified).

Dorsum of pronotum subcylindrical; lateral carinae absent. Prosternal process absent. Tympanum absent. Fully apterous. Hind femur slender, with lower basal lobe longer than upper. External apical spine of hind tibia present. Male cercus simple, narrow conical; supra-anal plate angular; subgenital plate short, subconical divided on separate sclerites connected by membrane.

Sound-producing mechanism (in male only) represented by convex ridges on sides of third abdominal tergite densely covered with fine, transverse ridgelets. Second part of mechanism represented by sharp ridge at base of inner side of hind femur. Sound produced by rubbing femur against abdominal ridge. (Sound is probably supersonic, because direct observation proved it was not audible for human ear).

PHALLIC COMPLEX: Ectophallus membraneous except slightly sclerotized dorsal part, which can be considered as structure analogous to rudimentary cingulum. Endophallus membraneous consisting not differentiated endophallic sac, which is ending with weak distal sclerotization. Epiphallus disc-shaped, poorly sclerotized, with membraneous middle part of disc, more sclerotized sides, with pair of large, protruding forwards, inside projections and two-three small teeth near inner margins.

Spermatheca simple, of oval shape main reservoir without diverticula and with simple, narrow spermathecal duct.

Karyotype: Not known.

Type genus: *Tanaocerus* Brunner, 1906.

Only two genera of this family are known — type genus and genus *Mohavacris* Rehn, 1948.

Distribution: South West part of North America.

Genus *Tanaocerus* when first described by Brunner was placed by him in *Eremobiinae*. Rehn (1948) when describing the second genus — *Mohavacris* erected for them subfamily *Tanaocerotinae* and placed it in the family *Eumastacidae*, Dirsh, 1956 raised the subfamily to family rank, removing it from *Eumastacidae*. At present it is considered as a family of the superfamily *Pneumoroidea*.

The structure of the phallic complex suggests that *Tanaoceridae* have remote affinity with the families *Xyronotidae* and *Pneumoridae*. The primitiveness and certain similarity in the phallic complex and a similar type of sound-producing mechanism indicate that such a possibility cannot be ignored.

Superfamily

Pamphagoidea

Diagnosis: PHALLIC COMPLEX: Ectophallus partly sclerotized, forming sclerotized capsule or semicapsule structure or differentiated cingulum. Endophallus with strongly sclerotized penis. Epiphallus of various shape, with oval sclerites and lateral appendages present or absent; ancorae and lophi mostly present. Spermatophore sac in dorsal or ventral position.

Tegmina and wings fully developed, shortened, lobiform or absent. Tympanum mostly present. Sound producing mechanism if present of various types. Antennae shorter than body.

List of subfamilies:

1. *Charilaidae*
2. *Lathiceridae*
3. *Ommexechidae*
4. *Pamphagidae*
5. *Pyrgomorphidae*

Key to families

1 (6)   Penis' sclerites divided on basal and apical valves.

2 (5)   Epiphallus shield-like shaped. Antennal grooves absent. Cingulum differentiated.

3 (4)   Valves of penis articulated. Epiphallus without ventro-lateral appendages or oval sclerites. Median carina of pronotum single.

Pamphagidae

4 (3)   Valves of penis not articulated. Epiphallus with ventro-lateral appendages. Median carina of pronotum double.

Charilaidae

5 (2) Epiphallus bridge-shaped. Antennal grooves present. Sclerotized ectophallus semicapsule-like, not forming cingulum.

Lathiceridae

6 (1) Penis' sclerites not divided on basal and apical valves.

7 (8) Epiphallus with dorso-lateral appendages; lateral plates fused with bridge. Spermatophore sac in dorsal position.

Pyrgomorphidae

8 (7) Epiphallus with oval sclerites; lateral plate connected with bridge by membrane. Spermatophore sac in ventral position.

Ommexechidae

Family

Pamphagidae

Diagnosis: Body from very large to medium size, dorso-ventrally depressed, laterally compressed or subcylindrical. Great sexual dimorphism often present. Integument mostly strongly rugose. Head subconical, conical or subglobular; face in profile excurved, straight or deeply incurved; frontal ridge present, mostly sulcate; fastigium of vertex of various shape, at apex with furrow; fastigial foveolae absent, but analogous concavities may present. Antennae ensiform, ribbon-like or rod-like. Dorsum of pronotum crested, tectiform, subcylindrical, or flat. Prosternal process or collar present. Mesosternal interspace open. Tympanum mostly present. Tegmina and wings fully developed, shortened, lobiform, rudimentary or absent; reticulation of tegmina dense; venation normal or highly specialised as stridulatory mechanism; wings normal or specialised for sound producing. Krauss' organ mostly present. Hind femora of various shape with lower basal lobe longer than upper. Brunner's organ present. External apical spine of hind tibia present or absent. Male cercus simple, conical; supra-anal plate angular; subgenital plate short subconical, conical or at apex bilobate. Ovipositor short, with curved valves.

Sound-producing mechanism, various and complicated, mostly present.

PHALLIC COMPLEX: Ectophallus membraneous or in distal half slightly sclerotized; cingulum well differentiated; valves of cingulum absent. Endophallus strongly sclerotized; basal valves of penis separated from apical valves, but articulated with them, and connected with rami; gonopore processes absent. (The similar protruding process from the basal valves is not functionally the same as gonopore process). Ejaculatory sac large, in ventral position. Spermatophore sac relatively small, in dorsal position. Phallotreme long, with ventral slit extending to ejaculatory sac. Epiphallus shield-like, with short ancorae and without lophi (in one subfamily secondary modified into bridge-like structure). Oval sclerites absent.

Spermatheca: Contains only main reservoir, sometimes curved or twisted and sometimes with lateral pockets, but lack of diverticula.
Karyotype: $2n\sigma = 19$.
Type genus: *Pamphagus*. Thunberg, 1815.

The family at present is divided into four subfamilies, which are clearly differentiated between themselves. They are:

1. *Akicerinae*
2. *Echinotropinae*
3. *Pamphaginae*
4. *Porthetinae*

Distribution: Palaearctic and Ethiopian Regions.

This homogenous family is so well defined and isolated from the other families of *Acridoidea* that it was always considered as a subfamily when superfamily *Acridoidea* was regarded as a family. At present it is accepted everywhere that *Pamphagidae* are a group of a family rank.

The only remote affinity of *Pamphagidae* is between them and family *Charilaidae*. Possible that they originated from the same ancestral stock, but branching took place very early.

Key to subfamilies

1 (2)   Middle tibia on dorsal side with row of teeth or tubercles. If winged, second vannal vein of wing S-like curved and first and third vannal area expanded.
Akicerinae

2 (1)   Middle tibia without teeth or tubercles. Wing venation not specialized.

3 (4)   Fastigium of vertex and upper part of frons far projecting forwards. Body elongated, narrow cylindrical. Epiphallus with deeply excised posterior margin.
Echinotropinae

4 (3)   Fastigium of vertex moderately or not at all projecting forwards. Body stout, subcylindrical, laterally compressed or dorso-ventrally depressed. Posterior margin of epiphallus not excised.

5 (6)   Costal area of tegmen expanded and covered with parallel, ridge-like veinlets. Apical valves of penis mostly serrated or teethed.
Porthetinae

6 (5)   Tegmina, if present, without specialization, mostly lobiform or absent. Apical valves of penis smooth.
Pamphaginae

Subfamily

Akicerinae

(Fig. 16)

Diagnosis: Body large or medium size, subcylindrical, dorso-ventrally depressed or compressed laterally. Sexual dimorphism mostly represented by difference in size. Integument mostly strongly rugose. Head from conical to subglobular; face, in profile, straight, slightly excurved or strongly incurved; frontal ridge sulcate; fastigium of vertex short or far protruding forwards; fastigial foveolae absent, but similar parts of sculpture sometimes present. Antennae filiform or ensiform. Dorsum of pronotum crested, tectiform or flat, mostly strongly sculptured. Prosternal process or collar present. Tympanum present. Tegmina and wings fully developed, shortened, vestigial or absent. Venation of tegmen not specialised. Second vannal vein of wing S-like curved and first and third vannal areas expanded. Krauss' organ present or absent. Middle tibiae, on dorsal side, with row of teeth or tubercles. Hind femora moderately widened and spiked. External apical spine of hind tibia mostly present.

Sound-producing mechanism represented by specialised venation of wings and, as a second part of mechanism, by serrated middle tibiae.

PHALLIC COMPLEX: Ectophallus membraneous except large, sclerotized cingulum with large zygoma, apodemes and rami; valves of cingulum absent. Edophallus

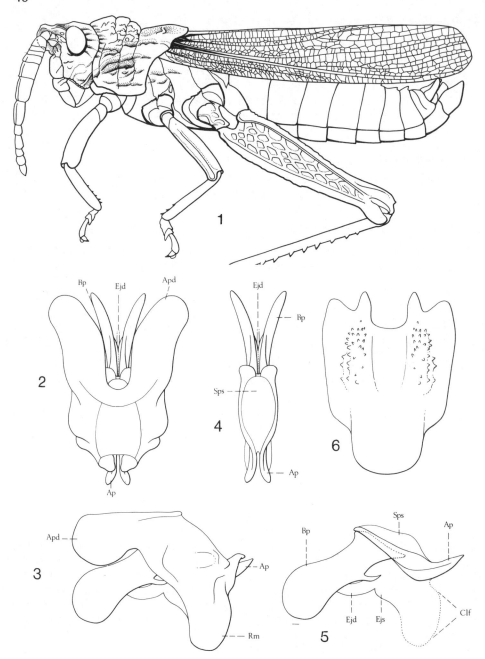

Figure 16.

Akicera fusca (Thunberg, 1815). 1, male. 2-6, phallic complex of *Strumiger desertorum*
Zubowsky, 1896. 2, phallic complex from above. 3, the same, lateral view. 4, endophallus
dorsal view. 5, the same, lateral view. 6, epiphallus.

strongly sclerotized; basal valves of penis moderately narrow and relatively long, on ventral side mostly with short, tooth-like process (not a gonopore process); articulation with apical valves of penis clear; apical valves of penis slightly upcurved with apices subacute or obtuse, with smooth surface. Epiphallus shield-like, sometimes longer than wide, sometimes vice-versa; two clusters of teeth or tubercles, laterally to middle of shield, present; ancorae well developed; lateral plates mostly well developed; lophi absent.

Spermatheca: Of simple structure, containing moderately wide apical main reservoir often curved or twisted, sometimes with pocket-like convexities. Diverticula absent.
Karyotype: $2n\,\male = 19$.
Type genus: *Akicera* Serville, 1831.

The subfamily contain about 30 genera and can be easily divided into four, well defined, tribes: *Akicerini*, *Adephagini*, *Thrinchini* and *Batrachotetrigini*.

Distribution: Ethiopian, Palaearctic and Western part of Oriental Region.

The subfamily was erected by Dirsh, 1961 as an assemblage of representatives of the various groups of the taxon connected mostly by the same type of the sound-producing mechanism.

Subfamily

Echinotropinae

(Fig. 17)

Diagnosis: Body strongly elongated, narrow cylindrical. Great sexual dimorphism present. Integument strongly rugose. Head conical; face, in profile, strongly incurved; frontal ridge sulcate; fastigium of vertex elongated and, in profile, strongly protruding forwards, at apex rounded or angular; fastigial foveolae absent. Antennae ensiform, triangular in cross section. Dorsum of pronotum of various shape, tuberculate and spiny. Prosternal process collar like. Tympanum present or absent. Tegmina and wings fully developed, lobiform or absent. Tegmen reticulation dense; venation of tegmina and wings not specialised. Hind femora slender. External apical spine of hind tibia present or absent. Krauss' organ absent.
Sound-producing mechanism not found.

PHALLIC COMPLEX: Ectophallus membraneous, in apical part slightly sclerotized, with strongly sclerotized cingulum; zygoma short, apodemes relatively long and wide; rami narrow. Endophallus strongly sclerotized; basal valves of penis wide, with almost truncate apices, closely articulated with apical valves, which are relatively narrow, in apical part partly fused with rami. Epiphallus with large posterior incision giving to the structure bridge-shaped appearance; pair of rows of teeth scattered along lateral plates.

Spermatheca: Unknown.
Type genus: *Echinotropis* Uvarov, 1944.

Four genera and a few species of this subfamily are known. The subfamily is strongly deviating from the other subfamilies of *Pamphagidae*.

Distribution: South and S.W. Africa.

As a subfamily *Echinotropinae* were first established by Dirsh, 1961.

By the shape of head *Echinotropinae* are in certain respects similar to *Porthetinae*, otherwise they have no other affinities with the latter subfamily or with any other subfamily of *Pamphagidae*.

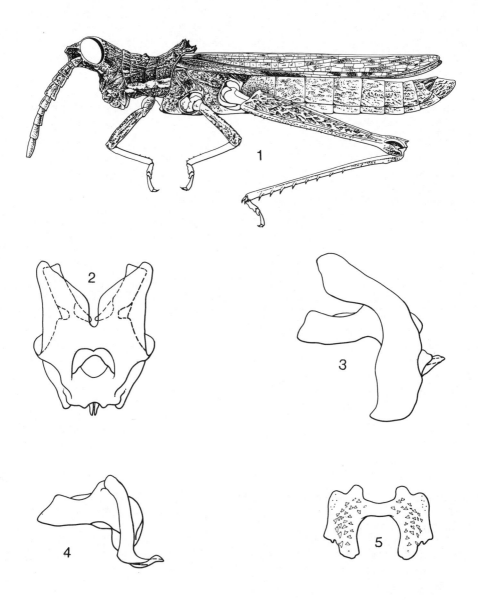

Figure 17.

Parageloiomimus spinosus Dirsh, 1956. 1, male. 2, phallic complex, dorsal view. 3, the same, lateral view. 4, endophallus, lateral view. 5, epiphallus. (Figs. 2-5, *Geloimimus nasicus* Saussure, 1899).

Subfamily

Porthetinae

(Fig. 18)

Diagnosis: Body large, laterally compressed or dorso-ventrally depressed. Great sexual dimorphism present. Integument rugose. Head conical or subconical; face, in profile, straight or concave at ocellus; frontal ridge sulcate; fastigium of vertex protruding forwards, at apex angular or rounded; fastigial foveolae absent. Antennae ensiform, differentiated or ribbon-like. Dorsum of pronotum crested, tectiform or flattened; lateral carinae absent. Prosternal process or collar present. Mesosternal interspace open. Tympanum present. Tegmina and wings fully developed, shortened or absent (in females); reticulation of tegmen dense; venation specialised for sound-producing (except species with lobiform tegmina). Hind femora wide or very wide, strongly spiked. External apical spine of hind tibia present. Krauss' organ present.

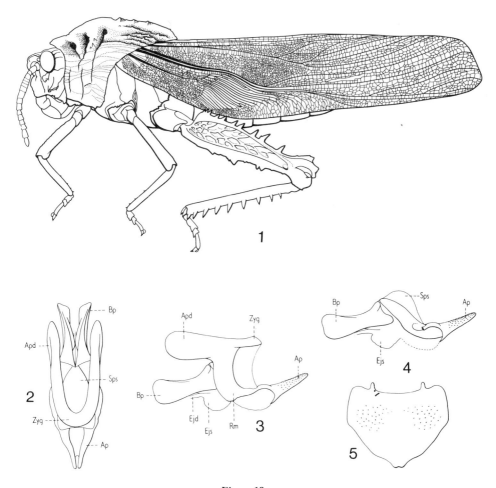

Figure 18.

1, *Porthetis carinatus* (Linnaeus, 1758). 2-5, *Cultrinotus luanensis* Uvarov, 1953, phallic complex. 2, dorsal view. 3, lateral view. 4, endophallus, lateral view. 5, epiphallus.

52

Sound-producing mechanism represented by expanded costal area of tegmen with row parallel sharp veinlets; the second part of the mechanism represented by tubercles or rough margins of inner side of hind femur. In wingless female sounds produced by rubbing hind femur against tuberculate side of abdomen.

PHALLIC COMPLEX: Ectophallus membraneous; except sclerotized U-shaped cingulum; zygoma short and narrow; apodemes relatively long. Endophallus strongly sclerotized; basal valves of penis narrow, articulation with apical valves of penis well expressed; apical valves of penis relatively long, in distal part covered with small teeth, or with serrated edges. Epiphallus shield-like; two clusters of teeth of various size present in middle of lateral sides of the shield; ancorae well developed; lateral plates form sides of shields.

Spermatheca: Contains large apical, downcurved reservoir and rather wide, twisted in distal part spermathecal duct.
Karyotype: 2n ♂ = 19.
Type genus: *Porthetis* Serville, 1831.

This subfamily, beside type genus, contains thirteen known genera.

Distribution: South, South West and South East Africa, South West part of Arabia.

Porthetinae were considered previously as a group *Portheti* (Dirsh, 1958). Later they were upgraded to subfamily rank (Dirsh, 1961) on the basis of the peculiar sound-producing mechanism characteristic for the taxon.

Subfamily

Pamphaginae

(Fig. 19)

Diagnosis: Body large to medium size, compressed laterally or dorso-ventrally depressed; sexual dimorphism is manifestating only in size of body. Integument mostly rugose. Head subconical; face, in profile, straight, slightly excurved or slightly incurved; frontal ridge sulcate; fastigium of vertex short, angular; fastigial foveolae absent. Antennae filiform, ribbon-like or slightly ensiform. Dorsum of pronotum crest-like, tectiform, subcylindrical or flattened. Prosternal process or collar present. Tympanum present or absent. Tegmina and wings lobiform, vestigial or absent. Hind femora from rather wide to comparatively slender. External apical spine of hind tibia present. Krauss' organ mostly present.

Sound-producing mechanism mostly present in winged genera and represented by venation of tegmen and wing which can produce sound by rubbing one against another.

PHALLIC COMPLEX: Ectophallus membraneous, except relatively large and strongly sclerotized cingulum, with zygoma, apodemes and strongly enlarged rami. Endophallus strongly sclerotized; basal valves of penis relatively wide, with ventral short, tooth-like projection; articulation with apical valves of penis located in dorso-distal ends of basal valves. Apical valves strongly sclerotized of various shape and

---

Figure 19. ▶

1, *Pamphagus elephas* (Linnaeus, 1758), male. 2-6, *Tropidauchen marginatum* I. Bolivar, 1912. Phallic complex: 2, dorsal view. 3, lateral view. 4, endophallus, dorsal view. 5, endophallus, lateral view. 6, epiphallus.

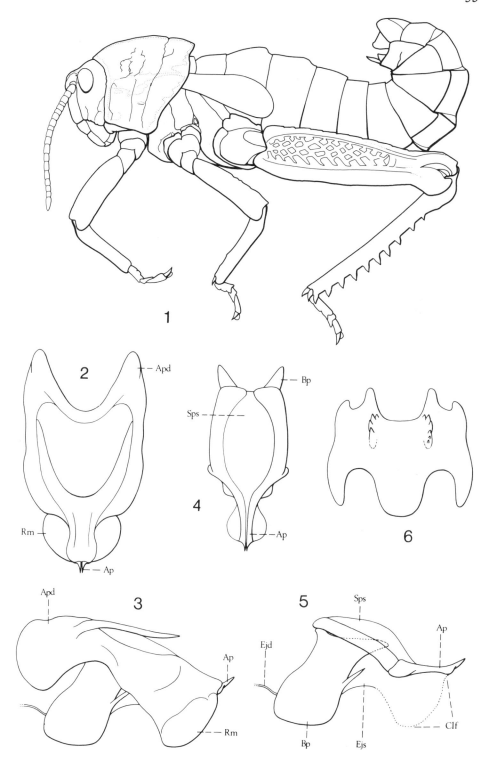

relatively long, at apex acute or subacute; in distal part smooth. Epiphallus shield-like of various shape; laterally to middle with two clusters of teeth, ridges, or tubercles; ancorae well developed; lateral plates well or poorly developed.

Spermatheca: Represented by single apical, rather twisted, moderately narrow main reservoir and twisted, widened in distal part spermathecal duct.
Karyotype: $2n\,\male = 19$.
Type genus: *Pamphagus* Thunberg, 1815.

The subfamily contains about 40 known genera. The subdivision of *Pamphaginae* on lower taxa is possible, but rather difficult, owing to apterism or micropterism of the subfamily.

Distribution: N. Africa, Mediterranean subregion, S.W. Asia and middle part of Asia up to Far East.

Interrelation of *Pamphaginae* with other subfamilies of the family may be established after studying the phallic complex of all genera, which is not done yet.

Family

Charilaidae

(Fig. 20)

Diagnosis: Body of medium size, subcylindrical. Integument slightly rugose. Head conical; face, in profile, incurved; frontal ridge sulcate; fastigium of vertex angular or parabolic; fastigial furrow present; fastigial foveolae or areolae absent. Antennae filiform. Dorsum of pronotum with pair of median carinae; lateral carinae present. Prosternal process present. Mesosternal interspace open, with furcal suture curved backwards. Tympanum present. Tegmina and wings fully developed, shortened or lobiform. Venation of tegmen and wing in macropterous forms specialized. Hind femora slender, with lower basal lobe longer or equal to upper one. External apical spine of hind tibia present. Male cercus straight, conical or slightly downcurved; supra-anal plate angular; subgenital plate short, subconical. Ovipositor wide, short and sturdy; valves curved at apices.
     Sound-producing mechanism of tegmen - wing type. In fully winged forms consists of thickened in basal part branches of medial veins of tegmen; on wings costal area in middle part expanded, with row transverse, strongly thickened veinlets; cubital and subcubital area strongly expanded, with sparse, regular transverse veinlets.

PHALLIC COMPLEX: Ectophallus in proximal part membraneous, in distal part sclerotized; cingulum or analogous structure present, and consists pair of dorso-lateral sclerites which can be considered as disjoined apodemes; zygoma and rami absent; valves analogous to valves of cingulum present. Endophallus strongly sclerotized; basal and apical valves of penis disconnected; basal valves large, robust, at proximal ends expanded; apical valves of penis relatively weak their apices covered by valves of cingulum. Ejaculatory sac in ventral position; spermatophore sac in dorsal position; phallotreme wide with ventral slit. Epiphallus shield-like, with large anterior projections and ventro-lateral appendices; ancorae and lophi absent. Oval sclerites absent.

Spermatheca: Consists relatively narrow, S-curved main reservoir and small diverticulum. Spermathecal duct narrow, widening at basal part.

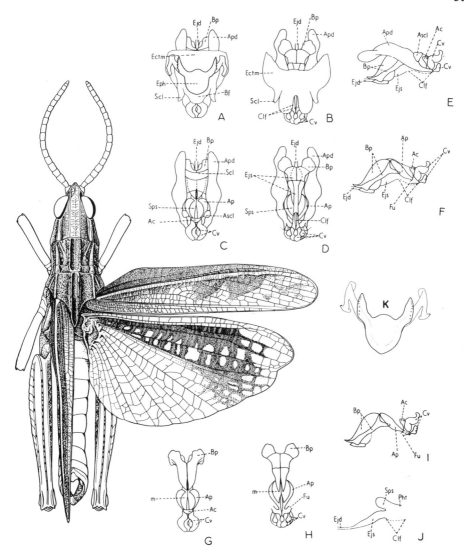

Figure 20.

Charilaus carinatus Stal, 1875. Male. Phallic complex. A, B, whole complex, dorsal and ventral views; C, D, E, phallic organ, epiphallus and ectophallic membrane removed, dorsal, ventral and lateral views; F, G, H, as for E, but apodemes and small sclerites (Ascl) removed, lateral, dorsal and ventral views; I, as for F, but sacs removed and only sclerotized parts left; J, endophallic sacs in profile, diagrammatic; K, epiphallus.

Karyotype: 2n♂ =23.
Type genus: *Charilaus* Stal, 1875.

  Beside the type genus, only three more genera of this family are known.

Distribution: Two widely disconnected area of this family are known. South and South West Africa (three genera) and North Africa (one genus).

Stal, 1875 describing genus *Charilaus* placed it into present *Pyrgomorphidae*. Karsch, 1896 suggested that it belongs neither to present *Pyrgomorphidae* nor to *Pamphagidae* and transferred it into group "*Caloptenoden*". Saussure, 1889 considered *Charilaus* as Pamphagid and so did I. Bolivar, 1916. Uvarov, 1943 expressed opinion that *Charilaus* belongs to the *Pamphaginae-Pyrgomorphinae* complex. Dirsh, 1953, on the basis of the structure of the phallic complex, erected for the group a new subfamily *Charilainae* and in 1956 raised it to a family rank.

The family is related to the family *Pamphagidae* owing to several common characters of phallic complex, such as divided penis' sclerites and dorsal position of spermatophore sac. From the other hand, *Charilaidae* differ strongly from *Pamphagidae* by peculiar and complicated ectophallus, with unusual cingulum and its valves, by structure of epiphallus and in external characters by double median carina of pronotum.

Probably, that *Charilaidae* are a relic of the earlier branch of the ancestral stock of both the families. The disconnected area of geographical distribution of *Charilaidae* supports a point of view of their more ancient origin than *Pamphagidae*.

Family

Lathiceridae

(Fig. 21)

Diagnosis: Body large or medium size, very robust and mostly dorso-ventrally depressed. Head subglobular, mostly prognathous; face, in profile, straight or excurved; fastigium of vertex short and wide; fastigial furrow present; fastigial foveolae absent; pair of deep grooves on sides of frontal ridge, into which antennae can be folded, present. Ocelli rudimentary or obliterated. Mandible extremely powerful. Antennae short, rod-like 7-13 segmented. Dorsum of pronotum wider than its length, flattened, carinae absent. Prosternal process present. Mesosternal interspace very wide, open. Tegmina, wings and tympanum absent. Hind femora short, wide and robust, with lower basal lobe longer than upper one. Brunner's organ present. External apical spine of hind tibia absent. Male cercus short, subconical; supra-anal plate angular; subgenital plate wide, subconical. Ovipositor short, robust with valves slightly curved.

Sound-producing mechanism not found.

PHALLIC COMPLEX: Ectophallus membraneous, except irregular supporting lateral sclerotization and strongly sclerotized, shield-like structure in proximal half of ectophallus. Endophallus strongly sclerotized; sclerites of penis divided; basal valves of penis relatively small, of irregular shape slightly widening towards proximal ends, gonopore processes absent, apical valves of penis relatively wide and sturdy, at apices covered with membraneous sheath; ejaculatory and spermatophore sacs poorly differentiated; ejaculatory sac in ventral position; spermatophore sac in middle position, between valves of penis; phallotreme short, with short ventral slit. Epiphallus bridge-shaped; bridge long and narrow; ancorae absent; lophi represented by pair, short, curved teeth. Oval sclerites present.

Spermatheca: Consists of apical comparatively narrow, S-curved main reservoir. Diverticula absent.
Karyotype: Unknown.
Type genus: *Lathicerus* Saussure, 1888.

Besides the type genus, three genera of this rare and peculiar family are known.

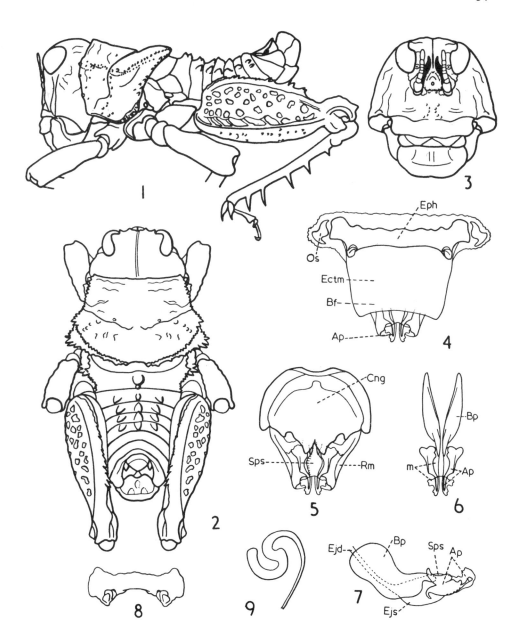

Figure 21.

1-3, *Batrachidacris tuberculata* (Rehn, 1956). 1, male, in profile. 2. The same, from above. 3. Face (antennal grooves painted black). 4-7. Phallic complex of *Batrachidacris rubridens* (Uv.). 4. Whole phallic complex from above. 5. The same, but ectophallic membrane and epiphallus removed. 6. Penis, from above. 7. Endophallus, in profile. 8. Epiphallus of *Batrachidacris tuberculata* (Rehn). 9. Spermatheca *Lathicerus cimex* Sauss.

Distribution: Known from South and South West Africa only.

The first two genera of the family *Lathicerus* and *Crypsicerus* were described by Saussure in 1888 and placed by him in the group "*Trincites*" of *Oedipodinae.* Uvarov, 1943 transferred the group to the tribe *Trinchini* of *Pamphagidae.* Dirsh, 1954, on the basis of the structure of the phallic complex, raised the group to subfamily and in 1956 to family rank.

The interrelation of the family with other families of the superfamily *Acridoidea* is not clèar. Their peculiar external appearance and characters, also peculiar phallic complex do not allow to link it closely with any of the known families. Their habit to burrow into soil is also peculiar and unparalled in the whole *Acridomorpha.*

Family
Pyrgomorphidae

Diagnosis: Body of various shape, from short and sturdy, cylindrical, fusiform, to narrow cylindrical — straw-like, from small to large size. Head conical, acutely conical or elongated conical. Fastigium of vertex of various form. Fastigial furrow present. Fastigial foveolae absent. Fastigial areolae mostly present. Antennae filiform, rod-like or ensiform. Dorsum of pronotum of various shape. Prosternal process, tubercle, or collar present. Mesosternal interspace of various form, open or closed. Tympanum present, sometimes rudimentary or absent. Tegmina and wings fully developed, reduced or absent. Lower basal lobe of hind femur longer or shorter than upper. External apical spine of hind tibia present or absent.

PHALLIC COMPLEX: Ectophallus capsule-like, strongly or partly sclerotized, not differentiated into structure homologous to cingulum of *Acrididae*, forming various secondary structures. Endophallus strongly sclerotized. Penis formed by pair sclerites, not divided on basal and apical valves. Gonopore processes absent. Spermatophore sac sclerotized, in dorsal position. Ejaculatory sac in ventral position; phallotreme large, often with slit in ventro-distal part. Epiphallus bridge- or disc-shaped, without or with rudimentary ancorae, lophi hook-shaped strongly sclerotized; dorso-lateral appendices attached to sides of bridge.

Spermatheca of various shape from simple apical reservoir to complicated vermicular structure with several diverticula or without diverticula.
Karyotype: 2n ♂ = 17, 19. In most of studied cases 2n ♂ = 19.

The family is divided here on following subfamilies containing about 127 genera.

List of subfamilies

   1. *Atractomorphinae*
   2. *Chrotogoninae*
   3. *Desmopterinae*
   4. *Dictyophorinae*
   5. *Fijipyrginae*
   6. *Geloiinae*
   7. *Nereniinae*
   8. *Omurinae*
   9. *Phymateinae*
  10. *Psednurinae*
  11. *Pyrgacrinae*
  12. *Pyrgomorphinae*
  13. *Zonocerinae*

Distribution: Tropical, Subtropical and Southern part of the temperate zones of all World.

The nomenclature and grouping of the family *Pyrgomorphidae* undergone too numerous changes in its status and grouping from Linnaean to present time.

Thunberg in 1815 erected genus *Phymateus*. Since then the taxon which is known now as *Pyrgomorphidae* was called by the various authors as *Phymatinae, Phymateini, Phymateinae,* until Walker in 1870 raised it into the family rank *Phymateidae*. This family was subjected to numerous nomenclature changes and regrouping. Then it was found that the group name was used as a family name *Phymatidae* in *Hemiptera Heteroptera* by Laporte, 1832. The changed name *Pyrgomorphini* was used by Brunner von Wattenwyl as a new group name for the taxon. It was used by most acridologists since then.

Dirsh, 1956 considered the group as a family and retained the name *Pyrgomorphidae*. Kevan (at al.), who extensively studied the family, in 1964 divided it into five "series" containing 29 tribes. In 1970 he divided it into ten "series" subdivided into 30 tribes.

Interrelation of the family with other families of *Acridoidea* is still rather obscure. From one hand the undivided sclerites of penis consisting pair of separate sclerites, and the dorsal position of the spermatophore sac, suggest affinity with the family *Lentulidae*. From the other hand the presence of fastigial furrow, hinting on the remote affinity with *Ommexechidae*.

Most probable that *Pyrgomorphidae* was branched very early from pre-acridoid stock. Their wide uninterrupted area of distribution in the tropical and subtropical zones of the world indicates that they originated in the warm geological period of the planet and their distribution was not interrupted by later cooling of climate.

Diversity of *Pyrgomorphidae* in external form of their body indicates that their evolution took considerable time.

Furthermore, their phallic complex is extremely stable in its general structure in the whole family and greatly variable in details of secondary character in almost every genus. This indicates that origin of the family was probably monophyletic and not associated closely with the roots of the origin of other families.

The classification of the family, on this reason is based mostly on the external characters. The grouping on the basis of phallic complex present difficulties owing to diversity in the secondary structures in almost every genus.

There are several genera such like *Tenuitarsus* and *Schultessia* which present author hesitates to attach to the definite subfamily, but in cases of mostly highly diversed genera it is inevitable.

Key to subfamilies

1 (16) Lower basal lobe of hind femur longer than upper.

2 (15) Prosternum with process or tubercle.

3 (12) Antenna below or on level with lateral ocelli.

4 (11) First abdominal tergite without dorsal gland.

5 (10) Upper part of frons, in profile, not excised.

6 (9) Pronotum strongly sculptured with spines, teeth, tubercles, irregular folds and furrows and granules.

7 (8) Pronotum covered with teeth, spines, tubercles or granules. Antennae mostly filiform. Tegmina and wings mostly fully developed.

Taphronotinae

60

8 (7) Pronotum with irregular folds and deep furrows. Antennae mostly rod-like. Tegmina and wings mostly shortened.

Dictyophorinae

9 (6) Pronotum smooth, finely rugose, or finely granulose.

Pyrgomorphinae

10 (5) Upper part of frons in profile excised.

Geloiinae

11 (4) First abdominal tergite with dorsal gland.

Zonocerinae

12 (3) Antennae placed in front of lateral ocelli.

13 (14) Body compressed laterally. Head short, acutely conical. Fully winged.

Desmopterinae

14 (13) Body depressed dorso-ventrally. Head elongated, narrow conical. Apterous or rarely winged.

Omurinae

15 (2) Prosternum with collar-like structure, covering mouth-parts below.

Chrotogoninae

16 (1) Lower basal lobe of hind femur shorter than upper.

17 (18) Mesosternal interspace open. Tympanum present. Antennae placed in front of lateral ocelli.

Atractomorphinae

18 (17) Mesosternal interspace closed. Tympanum absent. Antennae placed below lateral ocelli.

19 (22) Mouth parts not specialised.

20 (21) Dorso-lateral appendices separated from bridge of epiphallus.

Pyrgaerinae

21 (20) Dorso-lateral appendices not separated from bridge of epiphallus, of usual pyrgomorphoid type.

Nereniinae

22 (19) Mouth parts highly specialised.

23 (24) Almost fully winged (slightly reduced).

Fijipyrginae

24 (23) Apterous or micropterous.

Psednurinae

Subfamily

Taphronotinae

(Fig. 22)

Diagnosis: Body subcylindrical, from large, robust, to medium size. Head short, conical or elongated conical; face, in profile, incurved; fastigium of vertex from short to elongated at apex angular or rounded; fastigial areolae poorly developed. Antennae

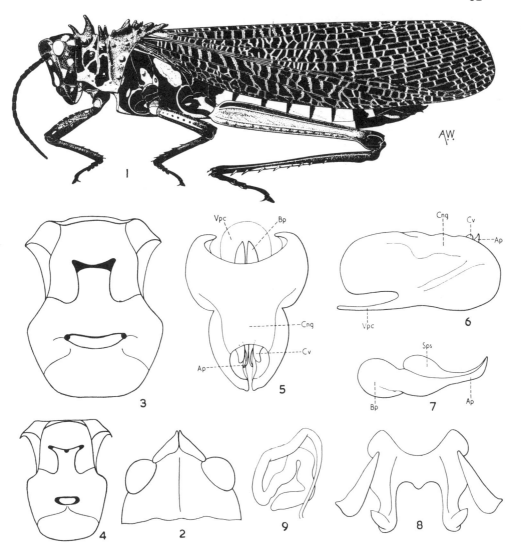

Figure 22.

Phymateus saxosus Coquerel, 1861. 1, female. 2, head from above. 3, meso- and metasternum, female. 4, the same, male. 5, phallic complex from above, epiphallus and ectophallic membrane removed. 6, the same, lateral view. 7, penis and spermatophore sac. 8, epiphallus. 9, spermatheca.

thick-filiform, with bases below lateral ocelli level. Dorsum of pronotum covered with teeth, tubercles or large granules, or combination of all of them. Prosternal process or collar-like elevation present. Mesosternal interspace open. Metasternal interspace distant from mesosternal. Tympanum present. Tegmina and wings fully developed or shortened; venation straight; reticulation dense. Hind femur narrow; with lower basal lobe longer than upper. External apical spine of hind tibia present. Male cercus short, conical; supra-anal plate angular; male subgenital plate short, subconical.

Sound-producing mechanism not detected, but some of them produce squeaking sounds.

PHALLIC COMPLEX: Ectophallus large, partly sclerotized, in distal part forming various, strongly sclerotized appendices. Endophallus relatively small, basal part of penis valves moderately expanded; distal part at apex curved upwards. Epiphallus bridge-shaped, with narrow bridge, without ancorae and with large, hooks ending lophi.

Spermatheca: In apical part simple, tube-like irregularly twisted, with single ending and sometimes with lateral, pocket-like bulging.
Karyotype: $2n\ \male = 19$.
Type genus: *Taphronota* Stal, 1873.

Few other genera necessary to mention to give a notion of the scope of the subfamily: *Paraphymateus* Dirsh, 1956; *Phymateus* Thumberg, 1815; *Aularches* Stal, 1873.

Distribution: Ethiopian, Malagassian and Oriental Regions.

Subfamily
Dictyophorinae

(Fig. 23)

Diagnosis: Body short, stout, subcylindrical, of large or medium size. Head conical, to acute conical; face, in profile, incurved; fastigium of vertex short, at apex rounded or obtuse angular; fastigial areolae poorly defined. Antennae short, rod-like of filiform, placed below lateral ocelli. Dorsum of pronotum irregularly subcylindrical, covered with numerous irregular wrinkles, furrows and irregular convexities. Prosternal process present. Mesosternal interspace open. Tympanum present, but sometimes rudimentary. Tegmina and wings with parchment-like membrane from fully developed to reduced or vestigial. Hind femora moderately slender, with lower basal lobe longer than upper one. External apical spine of hind tibia present. Male cerci, supra-anal and subgenital plate not specialized. Valves of ovipositor straight with obtuse apices.
Sound-producing mechanism not detected.

PHALLIC COMPLEX: Ectophallus large and partly sclerotized with strongly sclerotized apical appendices. Endophallus relatively large and strongly sclerotized; basal part of penis sclerites wide, with irregular edges; apical part relatively short, partly covered with sheath-like formation, at apex uncurved. Epiphallus bridge-shaped, with well defined ancorae and large, elongated, with large hooks, lophi; dorso-lateral appendices relatively large.

Spermatheca: Consists of numerous diverticula branched from apical part of spermathecal duct. Main reservoir not differentiated.
Karyotype: Unknown.
Type genus: *Dictyophorus* Thunberg, 1815.

Subfamily represented by few genera including *Maura* Stal, 1873; *Camoensia* I. Bolivar, 1881; *Parapetasia* I. Bolivar, 1884.

Distribution: Ethiopian Region.

The group *Dictyophorini* was erected by Kirby, 1902. Kevan, 1964, 1970 treated it as a tribe. Sum of characters, however, allow to consider them as subfamily.

Figure 23. ▶

1, *Dictyphorus spumans* (Thunberg, 1787). Female. 2-5, phallic complex. 2, dorsal view. 3, lateral view. 4, endophallus, lateral view. 5, epiphallus.

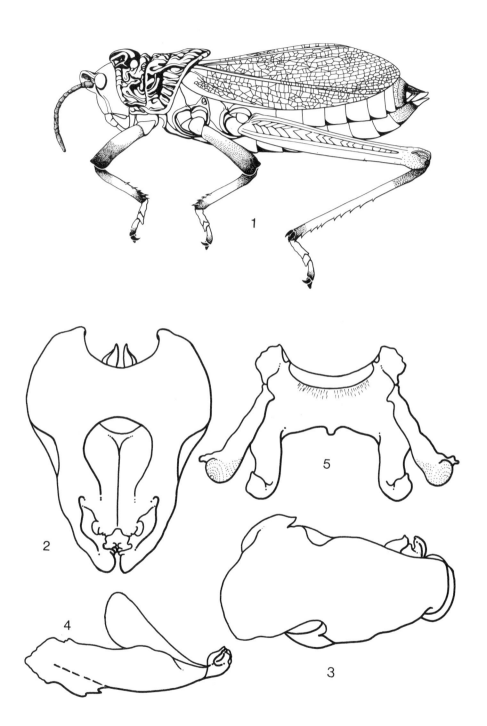

Subfamily

Pyrgomorphinae

(Fig. 24)

Diagnosis: Body of various shape, from small to large size. Head from conical to acutely conical; face, in profile, mostly incurved; fastigium of vertex from short to strongly elongated, at apex angular or rounded; fastigial areolae well developed, poorly defined, or completely absent. Antennae filiform, rod-like or narrow ensiform. Dorsum of pronotum of various shape. Prosternal process or tubercle present. Mesosternal interspace mostly open. Tympanum present or absent. Tegmina and wings fully developed, shortened, vestigial or absent. Venation usually simple, straight; reticulation mostly dense. Hind femur slender or moderately slender with lower basal lobe longer than upper one. External apical spine of hind tibia present or absent. Male cerci, supra-anal and subgenital plate mostly simple, unspecialized.

Sound-producing mechanism not detected.

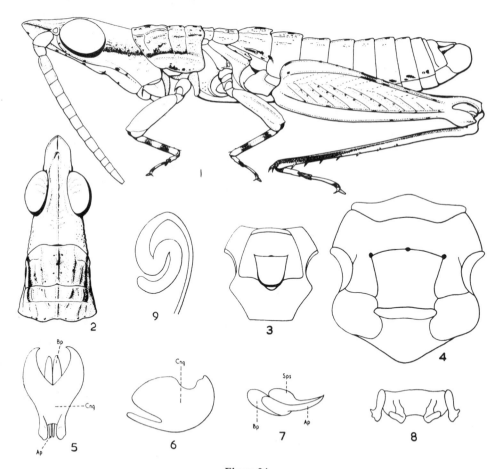

Figure 24.

1, *Pyrgomorphella minuta* Dirsh, 1963. Male. 2, head and pronotum from above. 3, meso- and metasternum, male. 4, the same, female. 5, phallic complex from above, (epiphallus and ectophallic membrane removed). 6, the same, lateral view. 7, penis and spermatophore sac. 8, epiphallus. 9, spermatheca.

PHALLIC COMPLEX: Ectophallus partly sclerotized, in distal part mostly forming sclerotized appendices, of various complecety. Endophallus much smaller than ectophallus; valves of penis strongly sclerotized, in basal parts in various degree expanded, in apical part usually slender, with apex acute. Epiphallus bridge-shaped, without ancorae; lophi of various relative size, mostly sturdy, ending with strong hooks.

Spermatheca: Mostly vermiform with single apical reservoir, sometimes with several vermiform diverticula.
Karyotype: 2n ♂ = 19.
Type genus: *Pyrgomorpha* Serville, 1838.

This subfamily is most numerous in its generic contents.

Distribution: Area of this subfamily is coincident with distribution of all family *Pyrgomorphidae.*

The history of the subfamily from nomenclature point of view is practically identical with the history of the family *Pyrgomorphidae.*
Kevan (1964, 1970) divided this subfamily into numerous tribes. Some of them deserving such merit but some of them are ambiguous. Since most of the genera are totally original in structure of secondary features, of the phallic complex but quite stable in the general plane of the primary characters of it, it is impossible to group them properly on the basis of this character. Grouping based on external characters is difficult owing to great diversity of them. Almost every genus in fact may be considered as a tribe, making useless classification of the group higher than generic level. But as a rather heterogeneous subfamily, comparable with other subfamilies of the family, *Pyrgomorphidae* represent useful systematic unit.

Subfamily

Geloiinae

(Fig. 25)

Diagnosis: Body subcylindrical or slightly fusiform, medium or large size. Head acutely conical; face, in profile, incurved; frontal ridge, between antennae, lamelliformly compressed, forming excision below apex of fastigium of vertex; fastigium of vertex elongate, at apex angular or narrow parabolic; fastigial areolae poorly developed. Antennae ensiform, rod-like, or filiform. Dorsum of pronotum cylindrical. Prosternal process or tubercle present. Mesosternal interspace open. Tympanum absent. Tegmina and wings absent or vestigial. Hind femora slender or moderately slender, with lower basal lobe longer than upper one. External apical spine of hind tibia present. Male cerci, supra-anal and subgenital plate from highly specialized to simple, unspecialized form.
Sound-producing mechanism not found.

PHALLIC COMPLEX: Ectophallus partly sclerotized, in distal end forming various shape sometimes very complicated, sclerotized appendices. Endophallus relatively large; basal part of penis sclerites strongly expanded and upcurved, apical part narrow, at apices acute. Epiphallus bridge-shaped, with narrow or wide bridge; ancorae absent; lophi large, strongly sclerotized, with short, strong hooks.

Spermateca: Represented by simple curved reservoir with narrow comparatively long spermathecal duct.

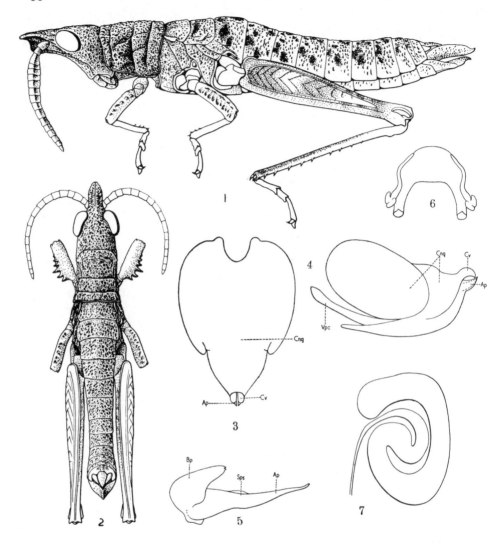

Figure 25.
Geloius finoti I. Bolivar, 1905. 1, female. 2, male. 3, phallic complex from above (epiphallus and ectophallic membrane removed). 4, the same, lateral view. 5, penis and spermatophore sac. 6, epiphallus. 7, spermatheca.

Karyotype: Unknown.
Type genus: *Geloius* Saussure, 1899.

Few other genera of the group exist — *Uhagonia* Saussure, 1905, *Pseudogeloius* Dirsh, 1963, *Pyrgohippus* Dirsh, 1963, *Saggitacris* Dirsh, 1963.

Distribution: Madagascar.

Saussure, when describing genus *Geloius* placed it into the "*Stripe Geloius*" of the tribe *Pyrgomorphii*. I. Bolivar (1905) elevated *Geloius* into subfamily *Geloiinae*. In 1909, however, he regarded it as a "Section *Geloii*". Since then several new genera of the group were described. Dirsh (1963) recognised them as "*Geloius* group". Kevan in 1964 and later in 1970 divided the group on several tribes (*Geloiini, Sagittacridini, Pseudogeloiini, Gymnohippini*). Here this group is treated in broader sense and is recognized as subfamily.

Subfamily

Zonocerinae

(Fig. 26)

Diagnosis: Body cylindrical or slightly fusiform, of medium size. Head conical or acutely conical; face, in profile, incurved; fastigium of vertex short, at apex angular; fastigial areolae indistinct. Antennae short, filiform. Dorsum of pronotum subcylindrical, slightly saddle-shaped or slightly flattened; lateral carinae absent. Prosternal process present. Mesosternal interspace open. First abdominal tergite with dorsal gland. Tympanum present. Tegmina and wings from fully developed to various degree shortened; venation straight, reticulation comparatively sparse. Hind femora relatively slender, with lower basal lobe longer than upper one. External apical spine of hind tibia present. Male cerci, supra-anal and subgenital plate not specialized.
Sound-producing mechanism not found.

PHALLIC COMPLEX: Ectophallus large and partly sclerotized, particularly in apical part. Endophallus relatively small; basal part of penis sclerites expanded, apical part slender, slightly curved, with acute apices. Epiphallus bridge-shaped, bridge narrow; ancorae absent; lophi short, stout, with strong, short hooks; lateral plates relatively large. Dorso-lateral appendices large.

Spermatheca: With twisted main reservoir and with very short diverticulum.
Karyotype: 2n $\male$ = 19.
Type genus: *Zonocerus* Stal, 1873.

Only two genera of this subfamily are known, the type genus and genus *Physemorphus* Krauss, 1907 endemic from Socotra Is.

Distribution: Ethiopian Region and Socotra Island.

Subfamily

Desmopterinae

Diagnosis: Body laterally compressed, of medium or large size. Head acutely conical; face, in profile, slightly incurved; fastigium of vertex short, at apex angular; fastigial areolae absent. Antennae ensiform or narrow ensiform, placed in front of lateral ocelli. Dorsum of pronotum flattened, with or without lateral carinae. Prosternal process present. Mesosternal interspace open. Tympanum present. Tegmina and wings fully developed or shortened; venation at tegmen straight, reticulation dense. Hind femora slender, with lower basal lobe longer than upper one. External apical spine of hind tibia present. Male cerci simple conical or incurved; supra-anal and subgenital plates not specialized.
Sound-producing mechanism not detected.

PHALLIC COMPLEX: Ectophallus partly sclerotized not forming distal appendices. Endophallus relatively large and sturdy; penis sclerites widened in middle; basal part

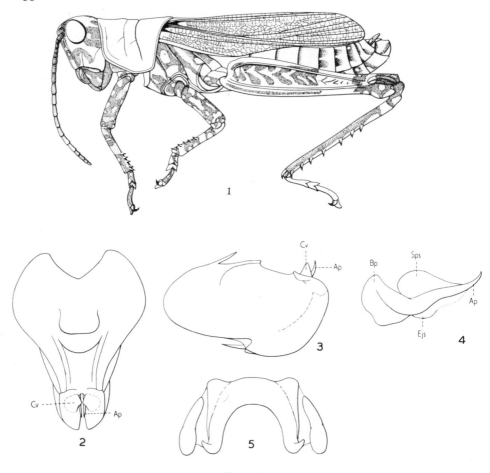

Figure 26.

1, *Zonocerus variegatus* (Linnaeus, 1758). Male. 2-5, phallic complex of *Zonocerus elegans* (Thunberg, 1815). 2, dorsal view, (epiphallus and ectophallic membrane removed). 3, the same, lateral view. 4, endophallus, lateral view. 5, epiphallus.

strongly expanded particularly upwards; apical part rather sturdy, with dorsal or apical appendix or with both of them. Epiphallus bridge-shaped; bridge comparatively narrow; ancorae absent, but apical projections sometimes closely resemble ancorae; lophi relatively short, sturdy, with short, strong hooks.

Spermatheca: Long, vermiform, irregularly twisted tube not widened at distal end, but sometimes slightly widened in middle.
Karyotype: Unknown.
Type genus: *Desmoptera* I. Bolivar, 1884.

Several genera of this subfamily are known.

Distribution: Austro-Asian archipelago, New Guinea and Australia.

Desmopterini as a group was first mentioned by I. Bolivar, 1905. Kevan, 1964, 1970, consider the group as a tribe. The combination of characters necessitate to elevate them to the subfamily rank.

Interrelation of the subfamily with the other subfamilies is rather uncertain, as is happening with most subfamilies of *Pyrgomorphidae*. Kevan, 1970 consider them as a rather isolated group in the family *Pyrgomorphidae*.

Subfamily

Omurinae

(Fig. 27)

Diagnosis: Body elongated dorso-ventrally depressed, of medium size. Head elongated narrow subconical; face, in profile, slightly incurved; fastigium of vertex strongly elongated, narrow angular; fastigial areolae indistinct. Antennae ensiform, with expanded first segment of flagellum, placed far in front of lateral ocelli. Dorsum of pronotum subcylindrical slightly flattened. Prosternal process present. Mesosternal interspace open. Tympanum absent. Tegmina and wings absent or strongly reduced. Hind femur slender, with lower basal lobe longer than upper one. External apical spine of hind tibia present. Male cerci, supra-anal and subgenital plate not specialized.
Sound-producing mechanism not found.

PHALLIC COMPLEX:Ectophallus relatively small, partly sclerotized, without specialized distal appendices. Endophallus relatively large, basal parts of penis sclerites strongly expanded; apical parts slender, not specialized. Epiphallus bridge-shaped; bridge narrow; ancorae absent; lophi elongated, with large strong hooks; apical projections very small.

Spermatheca: Not known.
Karyotype: Unknown.
Type genus: *Omura* Walker, 1870.

Distribution: South America.

Subfamily

Chrotogoninae

(Fig. 28)

Diagnosis: Body small, robust and mostly dorso-ventrally depressed, with integument tuberculate and spiny. Head short; face, in profile, incurved; fastigium of vertex short, at apex angular; fastigial areolae large, well defined. Antennae short, filiform, with apical segment elongated, truncheon-like. Dorsum of pronotum depressed, posterior angles of lateral lobes spreading sideways. Prosternal process collar-like covering lower mouth-parts. Mesosternal interspace open, wide; metasternal suture close to mesosternal interspace. Tympanum absent or rudimentary. Tegmina and wings fully developed, reduced or completely absent; membrane of tegmen parchment-like, venation straight, reticulation dense, irregular forming tubercles. Hind femora short, moderately wide; with lower basal lobe longer than upper one. External apical spine of hind tibia absent. Male cercus short, conical; supra-anal plate simple, angular; subgenital plate short, obtusely subconical.
Sound-producing mechanism not detected.

PHALLIC COMPLEX: Ectophallus large, slightly sclerotized, in distal end forming sclerotized appendices. Endophallus relatively small; proximal ends of penis sclerites

70

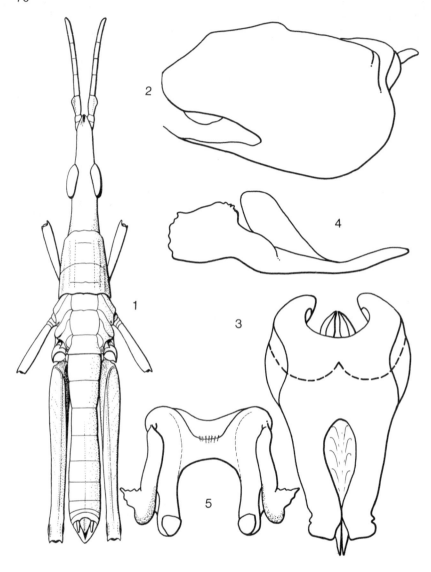

Figure 27.

1, *Omura congrua* Walker, 1870. Male. 2-5, phallic complex. 2, lateral view. 3, the same, dorsal view. 4, endophallus, lateral view. 5, epiphallus.

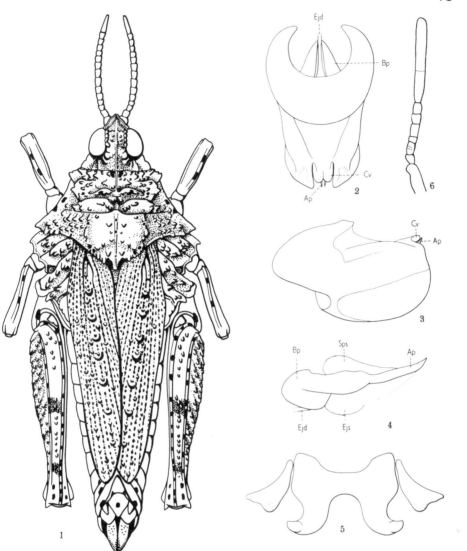

Figure 28.

1, *Chrotogonus homalodemus* (Blanchard, 1836). Female. 2-5, phallic complex of *Stibarosterna serrata* Uvarov, 1953. 2, phallic complex, dorsal view (membrane and epiphallus removed). 3, the same, lateral view. 4, endophallus, lateral view. 5, epiphallus. 6, antenna.

expanded, distal parts straight, with acute apices. Epiphallus bridge-shaped, with narrow bridge, ancorae absent; lophi short, sturdy, with strong hooks.

Spermatheca: Contents main narrow, twisted reservoir. Diverticula absent.
Karyotype: 2n ♂ = 19.
Type genus: *Chrotogonus* Serville, 1838.

Besides the type genus the subfamily contains four other genera.

Distribution: Ethiopian and Oriental Regions.

Chrotogoninae was first recognised as definite systematic unit — subtribe *Chrotogonae* I. Bolivar, 1884. Jacobson and Bianki, 1905 considered them as subfamily *Chrotogonini*. Since then they were frequently changing their status as "Subtribe", "Tribe", "Sectio", "Group", until Kevan in 1959 in his latest revision named them Tribe *Chrotogonini*. The combination of characters and a rather isolated position in relation to other *Pyrgomorphidae* make it necessary to elevate them to subfamily rank.

There is no other group *Pyrgomorphidae* that can be considered as near to the *Chrotogoninae*. The only character which unite it with the family is the phallic complex and presence of fastigial furrow.

Subfamily

Atractomorphinae

(Fig. 29)

Diagnosis: Body elongate subcylindrical, small or medium size. Head elongate, acutely conical; face, in profile, slightly incurved; fastigium of vertex elongate, angular; fastigial areolae obliterated or poorly developed. Antennae slightly ensiform, placed in front of lateral ocelli. Dorsum of pronotum subcylindrical. Tegmina and wings fully developed; venation straight, reticulation dense. Hind femora slender, with lower basal lobe shorter than upper one. Hind tibia with external apical spine. Male cerci, supra-anal and subgenital plate not specialized.

Sound-producing mechanism not found.

PHALLIC COMPLEX: Ectophallus partly sclerotized, with sclerotized distal appendices. Endophallus relatively small; basal part of penis sclerites strongly expanded, apical parts narrow, upcurved, with acute apices. Epiphallus shield-shaped, without ancorae and lophi.

Spermatheca: Long, with pair of vermiform diverticula.
Karyotype: 2n ♂ = 19.
Type genus: *Atractomorpha* Saussure, 1862.

Monogeneric subfamily.

Distribution: Tropical, subtropical and southern parts of temperate zones of Eastern Hemisphere.

The genus was elevated to tribe rank by I. Bolivar, 1884. Jacobson and Bianki (1905) considered it as subfamily. Later authors downgraded it again to the tribe rank in which status it has remained up to the present. However, the present author, according to combination of characters considers it as subfamily.

The subfamily has not close affinity with other subfamilies of *Pyrgomorphidae*.

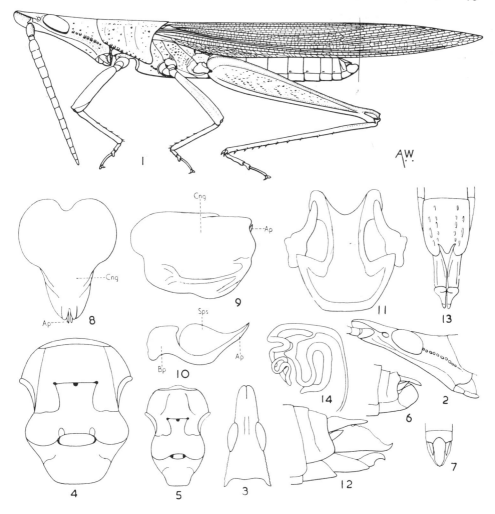

Figure 29.

Atractomorpha acutipennis (Guerin-Meneville, 1844). 1, male. 2, head of female, lateral view. 3, head of male from above. 4, meso- and metasternum, female. 5, the same, male. 6, end of male abdomen, lateral view. 7, the same, from above. 8, phallic complex from above, epiphallus and ectophallic membrane removed. 9, the same, lateral view. 10, penis and spermatophore sac. 11, epiphallus. 12, end of female abdomen, lateral view. 13, the same, from below. 14, spermatheca.

Subfamily

Pyrgacrinae

(Fig. 30)

Diagnosis: Body strongly elongated, narrow cylindrical, of medium size. Head elongate, narrow conical; opisthognathous; face, in profile, straight; fastigium of vertex elongated narrowing towards obtuse apex; fastigial foveolae or areolae absent. Galeae of maxillae protruding towards labrum. Antennae filliform, longer than head and pronotum together. Dorsum of pronotum cylindrical, without carinae. Prosternal

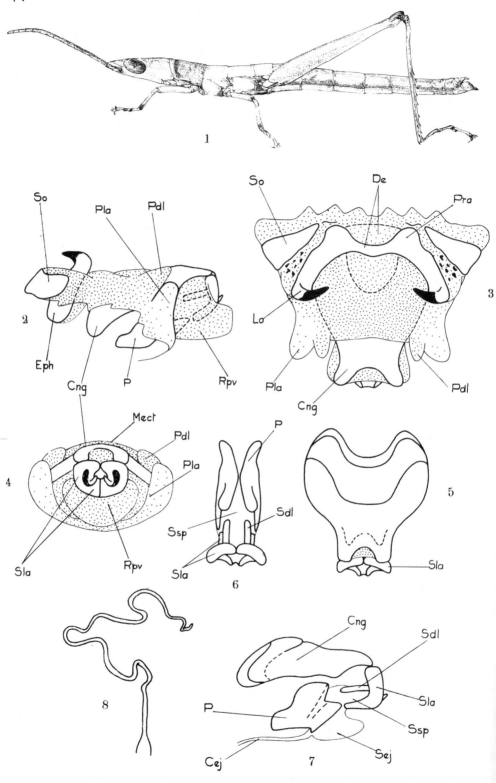

process present. Mesosternal interspace closed. Tympanum absent. Tegmina and wings absent. Hind femora slender, with lower basal lobe shorter than upper one. Hind tibia with external apical spine. Hind tarsus about half as long as tibia. Male cerci, supra-anal and subgenital plates not specialized.

Sound-producing mechanism not detected.

PHALLIC COMPLEX: Ectophallus capsule-like. Pyrgomorphoid, partly sclerotized, at distal end, in front of spermatophore sac complicated, strongly sclerotized structure, which can be confused as part of endophallus. Endophallus relatively small, with valves of penis short and very wide. Epiphallus bridge-shaped, with narrow bridge; ancorae absent; lophi narrow, with strong, incurved, acute hooks; lateral plates obliterated; dorso-lateral appendices separated from main body of epiphallus.

Spermatheca: Represented by long, narrow, vermicular, tube.
Karyotype: Unknown.
Type genus: *Pyrgacris* Descamps, 1968.

This unique subfamily contains only one genus and species.

Distribution: Known from Mauritius Island only.

This remarkable genus and species was described by Descamps, and its proper systematic position could not be found. It was thought that the genus possess a certain character of *Pyrgomorphidae* and *Hermiacridinae*. When my friend Dr. M. Descamps and myself personally discussed the matter I also was hesitating in what family and subfamily to place this genus. At present after long meditation I came to the conclusion that the apical parts of endophallus were interpreted incorrectly and if to consider them as highly specialized part of ectophallus then the genus is fit to be placed into *Pyrgomorphidae*. It is highly anomalous and probably is very ancient relic of the past fauna, and its relationship with other subfamilies at present cannot be suggested.

Subfamily

Nereniinae

(Fig. 31)

Diagnosis: Body of medium size, elongated, cylindrical. Head conical; face, in profile, incurved; fastigium of vertex short, at apex broadly angular; fastigial areolae poorly developed. Antennae filiform, comparatively long, exceeding length of head and pronotum together. Dorsum of pronotum subcylindrical. Prosternal process present. Mesosternal interspace closed. Tympanum absent. Fully apterous. Hind femora relatively slender, with lower basal lobe shorter than upper one. Hind tibia with external apical spine. Hind tarsus exceed half length of tibia. Male cercus, supra-anal and subgenital plate specialized forming complicated structures.

Sound-producing mechanism not found.

PHALLIC COMPLEX: Ectophallus large and partly sclerotized, in distal end forming sclerotized appendices. Endophallus relatively very large; sclerites of penis strongly

◀ **Figure 30.**

Pyrgacris relictus Descamps, 1968. 1, male type. 2-7, phallic complex. 2, lateral view. 3, dorsal view. 4, posterior view. 5, ectophallus, dorsal view (epiphallus and membraneous parts removed). 6, penis and distal parts of ectophallus (Sdl, Sla), dorsal view. 7, the same, lateral view. 8, spermatheca. (After Descamps, 1968. Lettering as in original drawing of Descamps).

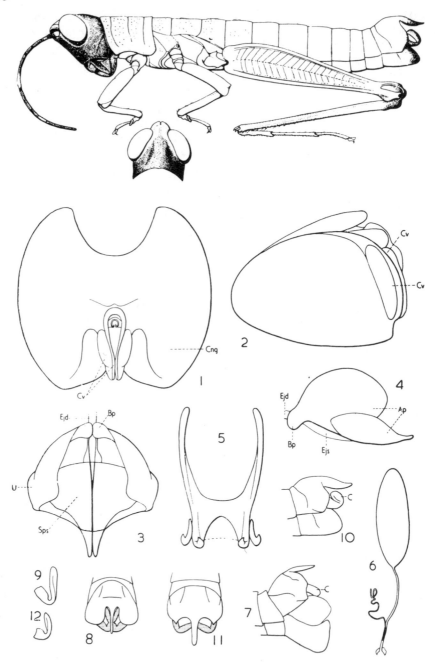

Figure 31.

Modernacris ysabelae Dirsh, 1964. Male type (whole insect and head). Phallic complex and spermatheca of *Modernacris controversa* Willemse, 1931. 1, phallic complex, from above (epiphallus removed); 2, the same lateral view; 3, endophallus, above; 4, the same lateral view; 5, epiphallus; 6, spermatheca; 7, end of male abdomen, lateral view (in erected position, the whole capsule of the cingulum being projected); 8, end of the male abdomen from above (cerci dotted); 9, left male cercus; 10-11, *Modernacris ysabelae* Dirsh, 1964; 10, end of the male abdomen, lateral view; 11, the same, from above (cerci dotted); 12, left male cercus.

expanded in middle, with small basal part; apical part strongly enlarged with upcurved outer sides. Epiphallus bridge-shaped; ancorae absent; anterior projections of lateral plates form long apodemes; lophi short, with strong, acute hooks.

Spermatheca: Consists very large, oval shaped main reservoir, short and narrow spermathecal duct, from basal part of which narrow vermicular diverticulum is branching.
Karyotype: Unknown.
Type genus: *Nerenia* I. Bolivar, 1905.

Only two genera *Nerenia* and *Modernacris* Willemse, 1931, are known..

Distribution: New Caledonia and Solomon Islands.

The genus *Modernacris* was described by Willemse (1931) and placed by him into *Cantantopinae*. Dirsh in 1964 after studying the phallic complex found that the genus belongs to the family *Pyrgomorphidae*.
Kevan, 1964 erected a separate tribe for these two genera which by the combination of characters are elevated here into subfamily rank.
It is possible that the *Nereniinae* are remotely related to the subfamily *Fijipyrginae*, sharing with this subfamily the structure of hind femora, closed mesosternal interspace, and elongated apodemes of anterior projections of epiphallus.

Subfamily

Fijipyrginae

Diagnosis: Body elongated cylindrical, of medium size. Head acutely conical; face, in profile, slightly incurved; fastigium of vertex moderately long, at apex narrow angular; fastigial areolae poorly developed, narrow. Antennae filiform, far exceeding length of head and pronotum together. Galeae of maxillae turned forwards, slightly overlapping margin of labrum; left galea smaller than right, reaching but not overlapping labrum. Dorsum of pronotum subcylindrical; lateral carinae absent. Prosternal process present. Mesosternal interspace closed. Tympanum absent. Tegmina and wings present, reaching ninth abdominal tergite. Hind femora slender, with lower basal lobe shorter than upper one. Hind tibia with external apical spine. Hind tarsus of half length of tibia. Male cerci, supra-anal and subgenital plate specialized.
Sound-producing mechanism not found.

PHALLIC COMPLEX: Ectophallus large, partly sclerotized not forming at distal end sclerotized appendices. Endophallus relatively large; basal parts of penis sclerites moderately expanded; apical parts narrow, at apex acute. Epiphallus disc-shaped; ancorae absent; anterior projections form long apodemes; lophi short ending with small, acute hooks.

Spermatheca: Female unknown.
Karyotype: Unknown.
Type genus: *Fijipyrgus* Kevan, 1966.

Only one genus and one species of this subfamily is known.

Distribution: Fiji Islands.

This subfamily was considered by Kevan, 1966 as a tribe of *Pyrgomorphidae*. The combination of external characters and the structure of the phallic complex suggests, however, that the appropriate taxon could be no less than of subfamily rank.

The interrelation of the subfamily with other subfamilies is obscure. The remote possibility is a certain affinity with the subfamily *Nereniinae.*

Subfamily

Psednurinae

Diagnosis: Body strongly elongated, straw-like, narrow cylindrical from medium to large size. Head elongated, narrow conical; face, in profile, straight or slightly incurved; fastigium of vertex elongate, narrow angular, at apex angular; fastigial areolae absent. Antennae narrow ensiform. Mouth parts specialized, galeae of maxillae turned upwards, reaching labrum. Dorsum of pronotum cylindrical, lateral carinae absent. Prosternal process present. Mesosternal interspace closed. Tympanum absent. Apterous or Micropterous. Hind femora slender, with lower basal lobe slightly longer than upper one. External apical spine of hind tibia present. Male cercus straight, conical or incurved, supra-anal plate angular, not specialized, subgenital plate strongly elongated, ensiform.

Sound-producing mechanism not found.

PHALLIC COMPLEX: Ectophallus relatively small and partly sclerotized, with tendency to form structure analogous to cingulum in *Acrididae*; distal part of ectophallus not specialized. Endophallus relatively large; sclerites of penis strongly sclerotized, basal parts expanded, apical parts narrow, slender, with acute apices. Epiphallus bridge-shaped; bridge narrow; ancorae absent; lophi long, stout ending with short, strong hooks.

Spermatheca: Represented by small, apical, irregular form reservoir and long, narrow, irregularly twisted spermathecal duct.
Karyotype: $2n\,\male = 19$.
Type genus: *Psednura* Burr, 1903.

This subfamily contains only two genera — *Psednura* Burr, 1907 and *Propsednura* Rehn, 1953.

Distribution: Australia and Tasmania.

The genus *Psednura,* when first described, was placed into subfamily *Psednurinae* of the family *Eumastacidae.* Rehn in 1953 transferred it into subfamily *Pyrgomorphinae.* Kevan, 1970 treated this group as a tribe. Several characters, particularly the peculiar specialization of mouth-parts, make it necessary to elevate this tribe into subfamily rank.

The affinity of this subfamily with other subfamilies of *Pyrgomorphidae* is rather obscure. Kevan (1970) considered a remote affinity with his tribes *Brunniellini* and *Verduliini.*

Family

Ommexechidae

(Fig. 32)

Diagnosis: Body large or medium size, subcylindrical or dorso-ventrally depressed. Integument mostly rugose. Head of variable shape; face, in profile, incurved or almost straight; frontal ridge sulcate; fastigium of vertex mostly angular; fastigial furrow present; fastigial foveolae and areolae absent. Antennae filiform. Dorsum of pronotum mostly flattened; lateral carinae irregular or indistinct. Prosternal process or tubercle present. Mesosternal interspace open. Tympanum present or absent. Tegmina and wings fully developed, shortened or absent; tegmen's membrane parchment-like; reticulation dense and irregular; venation of tegmen simplified with cubital vein unbranched. Hind femora relatively short, with lower basal lobe as long as or slightly longer than upper one. Lobes of hind knee obtuse angular. External apical spine of hind tibia absent. Male cercus simple conical; supra-anal plate angular; subgenital plate short, subconical. Ovipositor short with valves slightly curved and acute at apices.

Primitive tegmino-femoral sound-producing mechanism present or absent.

PHALLIC COMPLEX: Ectophallus membraneous except slightly sclerotized sheath covering apical part of endophallus and sclerotized cingulum; zygoma of cingulum elongate, narrow, apodemes short and slender, valves of cingulum absent. Endophallus relatively large; sclerites of penis not divided, but with middle part elongated and very thin; basal part of penis sclerites widened at proximal ends and slightly diverging; gonopore processes absent, apical part of penis sclerites strongly widened at apex, downcurved with serrated lower margin. Ejaculatory and spermatophore sacs unusually large, both in ventral position. Epiphallus bridge-shaped, bridge variate from wide to narrow; ancorae absent or hardly discernable; lophi tooth-like; lateral plates relatively large, connected with bridge by membrane; anterior projections from moderately protruding to almost obliterated; oval sclerites present.

Spermatheca: Spermathecal duct short, widening towards distal end, with several, of irregular form, diverticula.
Karyotype: $2n\,\male = 23$.
Type genus: *Ommexecha* Serville, 1831.

This family is not divided into subfamilies and contains besides the type genus five known genera.

Distribution: South America.

The type genus *Ommexecha* was first erected by Serville in 1831. Since then it was placed by various authors into group *Oedipodae* or *Pyrgomorphae*.

In 1884 I. Bolivar assigned for the group *Ommexecha* of subtribal status in the tribe *Pyrgomorphae*.

Kirby (1910) considered *Ommexechinae* as a subfamily of the family *Locustidae* (=*Acrididae*).

Dirsh (1956) raised the subfamily to family rank.

Eades (1961) again lowered the group to subfamily rank and divided it into three tribes — *Ommexechini, Aucacrini* and *Conometopini*. This addition of the two latter tribes to the subfamily *Ommexechinae* (sensu Eades) is supported only by the opinion of the mentioned author and not confirmed with the morphological data.

Dirsh (1961) considered *Ommexechidae* as a family and at present continue to consider it as such not dividing it into lower taxa and not adding new taxa.

The interrelation of the family with other families is not clear, from one hand the presence of fastigial furrow, shape of hind femora suggest remote affinity with *Pyrgomorphidae*, from the other hand similarity of the structure of epiphallus may suggest even more remote affinity with *Pauliniidae*. However, in both cases, particularly in latter one, it may be manifestation of convergent evolutionary development.

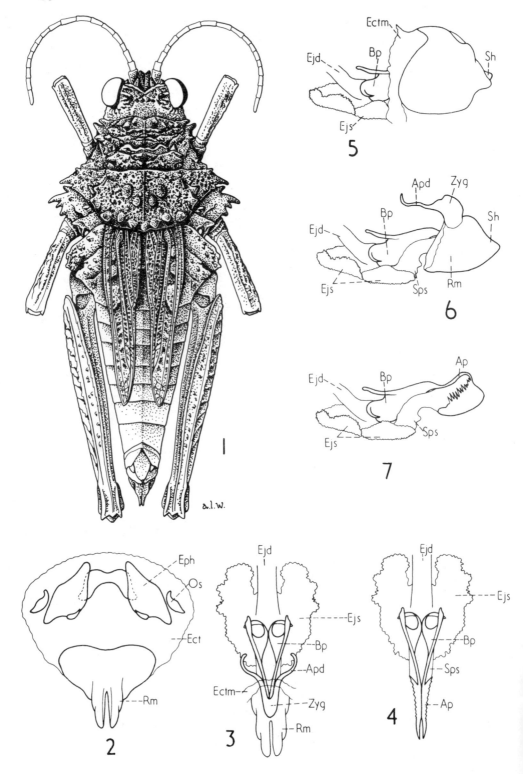

Superfamily

Acridoidea

Diagnosis: Body of various shape. Antennae shorter than body. Tegmina and wings fully developed, shortened, lobiform, vestigial or absent. Tympanum present or absent. Lower basal lobe of hind femur shorter or as long as upper one.

Sound-producing mechanism of various types; present or not detected.

PHALLIC COMPLEX: Ectophallus membraneous or partly sclerotized; cingulum strongly sclerotized and well differentiated. Endophallus with strongly sclerotized penis: penis sclerite not divided or divided on basal and apical valves which are fully separated or connected by flexure. Epiphallus of various shape, mostly with ancorae and lophi. Oval sclerites present.

List of families

1. *Acrididae*
2. *Catantopidae*
3. *Hemiacrididae*
4. *Lentulidae*
5. *Pauliniidae*

For difference between superfamilies *Acridoidea* and *Pneumoroidea* see the latter superfamily.

Key to families

1 (2)   Penis' sclerites not divided on basal and apical valves. Spermatophore sac in dorsal position.

Lentulidae

2 (1)   Penis sclerites divided on basal and apical valves, fully disconnected or connected by flexure. Spermatophore sac in middle or ventral position.

3 (6)   Basal and apical valves of penis sclerites fully disconnected.

4 (5)   Lateral plates of epiphallus connected with bridge by membrane.

Pauliniidae

5 (4)   Epiphallus bridge-shaped, with lateral plates fused with bridge.

Hemiacrididae

6 (3)   Valves of penis sclerites connected by flexure.

7 (8)   Ancorae of epiphallus not articulated with bridge. Sound-producing mechanism, if present, of various types.

Catantopidae

8 (7)   Ancorae of epiphallus articulated with bridge. Sound-producing mechanism, if present, of tegmino-femoral type.

Acrididae

◄

Figure 32.

1-7, *Ommexecha servillei* Blanchard, 1836. 1. Female. 2-7 Phallic complex. 2. Whole phallic complex from above. 3. The same, but ectophallic membrane and epiphallus removed. 4. Endophallus from above. 5. The same, in profile. 6. The same, with ectophallic membrane removed. 7. Endophallus in profile.

Family

Lentulidae

(Fig. 33)

Diagnosis: Body from small to medium size, of various shape. Integument from smooth to strongly rugose. Head of various shape; fastigial furrow absent; fastigial foveolae absent. Antennae filiform or ensiform. Dorsum of pronotum of various shape. Prosternal process or tubercle present. Mesosternal interspace open, rarely closed. Tegmina, wings and tympanum absent. Hind femora from slender to rather wide; lower basal lobe shorter than upper one; Brunner's organ present; lobes of hind

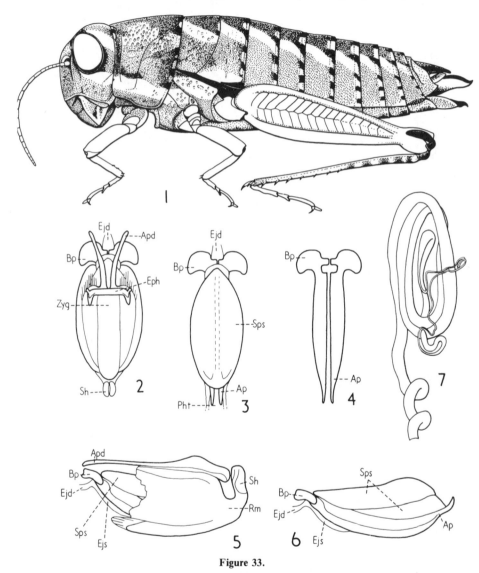

Figure 33.

Lentula callani Dirsh, 1956. 1. Male. 2-6. *L. obtusifrons* St. phallic complex. 2. From above. 3. From below. 4. Penis from above. 5. Whole phallic complex, in profile. 6. Endophallus, in profile. 7. Spermatheca.

knee rounded or obtuse angular. External apical spine of hind tibia present or absent. Male cercus, supra-anal plate and subgenital plate of various shape. Ovipositor of various shape.

Sound-producing mechanism not found.

PHALLIC COMPLEX: Ectophallus in proximal part membraneous, in distal part sclerotized, of semi-capsular structure, covering whole distal part of endophallus including apex of penis; cingulum with zygoma, apodemes and rami present; valves of cingulum absent. Endophallus strongly sclerotized; sclerites of penis not divided on basal and apical valves; basal part of penis sclerites widened, sometimes expanded; gonopore processes absent; ejaculatory sac small, in ventral position; spermatophore sac relatively large, in dorsal position; phallotreme long and wide, with long or short ventral slit. Epiphallus bridge-shaped; bridge relatively long and narrow; ancorae short, finger-shaped; lophi hook-shaped. Oval sclerites present.

Spermatheca: Main reservoir small, S-curved; no diverticula present. Spermathecal duct long, in basal half very wide, in apical half narrow.

Karyotype: 2n σ = 19, 20, 21, 23, 24. Predominantly 2n σ = 23.

Type genus: *Lentula* Stal, 1878.

The family contains more than twenty genera of a very diverse appearance.

Distribution: South, East and Central Africa.

Family *Lentulidae* was erected by Dirsh, 1956 on a basis of the peculiar phallic complex and nymphal appearance of the adult forms. The latter characters may be result of neoteny or degeneration as adaptation to certain ecological condition in remote past. At any rate the loss of tegmina and wings was accomplished relatively long ago, because indirect flight muscles are completely absent in *Lentulidae* (Ewer, 1958), but present in other studied wingless *Acridoidea*.

The interrelation of *Lentulidae* with other families of *Acridoidea* is obscure. In 1956 Dirsh expressed the opinion that *Lentulidae* may be remotely related to *Pyrgomorphidae*, however, since then the phallic complex of many other genera were studied, which indicates that such a possibility is hardly probable.

At present, after Brown (1967) published review of *Shelfordites*, this genus and allied newly described genera are removed from the family *Lentulidae* and form separate subfamily (see *Shelforditinae*), and transferred into family *Catantopidae*.

Family
Pauliniidae

(Fig. 34)

Diagnosis: Body of medium size or small, subcylindrical. Integument smooth. Head subconical; face, in profile, slightly incurved; fastigium of vertex short, at apex truncate or angular, without fastigial furrow; fastigial foveolae absent. Antennae filiform. Ocelli comparatively very large. Dorsum of pronotum flattened without lateral carinae. Prosternal process absent. Mesosternal interspace open. Tympanum present. Tegmina and wings fully developed or shortened; medial and cubital veins of tegmen sometimes unbranched. Hind femora moderately slender, with lower basal lobe shorter than upper one. Brunner's organ present. Lower lobe of hind knee angular or acute, spine-like. Hind tibia strongly expanded towards apex; external apical spine of hind tibia present or absent. Metatarsus of hind leg tarsi expanded. Male cercus acutely, conical, supra-anal plate angular; subgenital plate elongate, at apex bilobate.

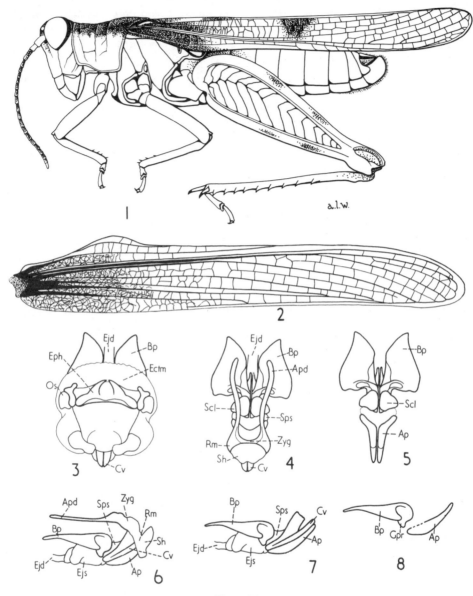

Figure 34.

Paulinia acuminata (DeGeer, 1773). 1. Male. 2. Right elyton. 3-8. Phallic complex. 3. Whole phallic complex from above. 4. The same, but ectophallic membrane and epiphallus removed. 5. Endophallus, from above. 6. As 4, but in profile. 7. Endophallus, in profile. 8. Penis, in profile.

Ovipositor shortened, with valves straight, hardly reaching apex of subgenital plate. Sound-producing mechanism not found.

PHALLIC COMPLEX: Ectophallus membraneous, except pair of ventral, distal, sclerotized plates, slightly sclerotized sheath covering apical part of penis; cingulum well differentiated, on zygoma, rami and apodemes; valves of cingulum present. Endophallus strongly sclerotized; penis' sclerites fully separated; into basal and apical valves; basal valves strongly expanded sideways; gonopore process present; apical valves of penis relatively small; ejaculatory sac very large, with pair of lateral pockets in basal part, placed ventrally to penis; spermatophore sac small, placed between and slightly above valves of penis. Epiphallus bridge-shaped; bridge moderately long; ancorae large, angular, slightly incurved, with acute apices; lophi relatively large, finger-shaped, upcurved; lateral plates moderately large attached to bridge by membraneous connections. Oval sclerites present.

Spermatheca: Main reservoir relatively small, oval; small diverticulum present; spermathecal duct in middle part widened.
Karyotype: 2n σ = 23.
Type genus: *Paulinia* Blanchard, 1843.

Besides the type genus of the family, second genus — *Marellia* Uvarov, 1929 is known.

Distribution: South America, Trinidad. (Possibly introduced recently for biological control of water weeds to South Africa).

Affinity of the family with other families of *Acridoidea* is doubtful. From one hand, the epiphallus with lateral plates and their membraneous articulation with the bridge are similar to that in *Ommexechidae*, but they have no other common characters. Possibly this common character is a result of independent development. From the other hand the absence of fastigial furrow, divided sclerites of penis and well differentiated cingulum *Pauliniidae* are sharing with *Acridinae* and *Hemiacridinae*, but again it could be attributed to the convergent and independent development of these characters in the groups which are not closely related. They possess the unique habit of semi-aquatic life and unique habit of sticking their egg pods on the lower surface of the water plants. Morphological changes facilitating life on the water surface and in the water itself and water plants, such as expanded tibiae and inflated metatarsus of hind legs indicate that adaptation was continuous for a very long time. This is obscuring the interrelation of the family even more.
The representatives of the family can swim and even submerge under the water. They have nocturnal habits.

Family
Hemiacrididae

Diagnosis: Body of various size and shape. Head of various shape. Fastigial furrow absent; fastigial foveolae or analogous structures present or absent. Antennae of various shape. Dorsum of pronotum of various form; carinae present or absent. Prosternal process or tubercle present. Mesosternal interspace open or closed. Tympanum present or absent. Tegmina and wings fully developed, shortened, lobiform, vestigial or absent. Lower basal lobe of hind femur shorter than upper one.
Sound-producing mechanism, if present, mostly of tegmino-tibial type.

PHALLIC COMPLEX: Ectophallus membraneous or partly sclerotized, cingulum present, well differentiated; valves of cingulum present or absent. Endophallus with penis' sclerites completely divided on basal and apical valves. Gonopore process present or absent. Epiphallus bridge-shaped, with or without ancorae; lophi present. Oval sclerites present.

Spermatheca: Of various shape.
Karyotype: Unknown.
Type genus: *Hemiacris* Walker, 1870.

Distribution: Tropical and partly subtropical zones of Whole World.

List of subfamilies

1. *Atacamacrinae*
2. *Chilacrinae*
3. *Conophyminae*
4. *Hemiacridinae*
5. *Leptacrinae*
6. *Leptisminae*
7. *Lithidiinae*
8. *Spathosterninae*

Key to subfamilies

1 (8) Valves of cingulum present. Tympanum present (rarely absent).

2 (7) Tegmina and wings fully developed or shortened. Radial area of tegmen with row of parallel transverse stridulatory veinlets.

3 (6) Antennae filiform. Body moderately long.

4 (5) Integument rugose. Head with incurved face or subglobular.

Hemiacridinae

5 (4) Integument smooth. Head subconical.

Spathosterninae

6 (3) Antennae mostly ensiform. Body strongly elongated stick-like.

Leptacrinae

7 (2) Fully apterous. Tympanum rudimentary or absent.

Conophyminae

8 (1) Valves of cingulum absent. Tympanum present or absent.

9 (12) Basal valves of penis forming long distal, dorsal projections.

10 (11) Antennae filiform. Tympanum present or absent (in apterous species).

Chilacrinae

11 (10) Antennae with large scape and elongated truncheon-like apical segment. Tympanum absent.

Atacamacrinae

12 (9) Basal valves of penis not forming distal dorsal projections.

13 (14) Fully winged. Tympanum present.

Leptysminae

14 (13) Fully apterous. Tympanum absent.

Lithidiinae

Subfamily

Hemiacridinae

(Fig. 35)

Diagnosis: Body from very large to small size, of various shape. Head prognathous or orthognathous; frons from excurved to strongly incurved; frontal ridge shallowly

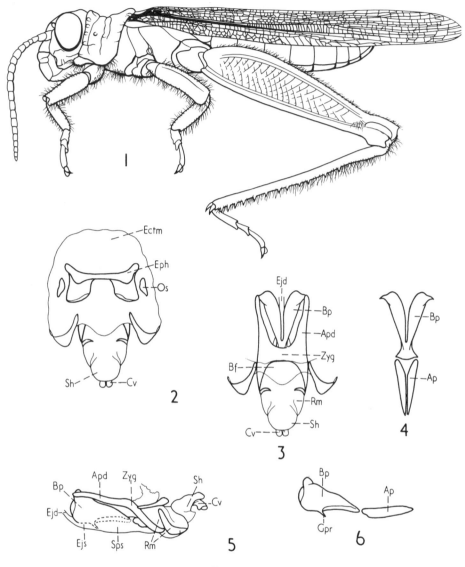

Figure 35.

Hemiacris fervens Walker, 1870. 1. Male. 2-6. Phallic complex. 2. Whole phallic complex from above. 3. The same, but ectophallic membrane and epiphallus removed. 4. Penis, from above. 5. As the fig. 3, but in profile. 6. Penis, in profile.

sulcate or obliterated; fastigium of vertex of various shape; fastigial foveolae or analogous structures present or absent. Antennae filiform. Dorsum of pronotum of various shape with lateral carinae present or absent. Prosternal process or tubercle present. Mesosternal interspace open. Tympanum fully developed, rudimentary or absent. Tegmina and wings fully developed, shortened or absent; reticulation of tegmen dense; radial area of tegmen with row thickened, regular, transverse veinlets (stridulatory specialization). Hind femora of various shape, mostly moderately wide, with lower basal lobe shorter than upper; lobes of hind knee short, angular or rounded. External apical spine of hind tibia mostly present. Male cercus from simple conical to very complicated shape; supra-anal plate angular; subgenital plate short, subconical or moderately short, narrow conical. Ovipositor short or moderately long with valves curved and acute at apices.

Sound-producing mechanism consists of stridulatory veinlets in radial area of tegmen. Stridulation probably performed by movement of hind tibia against the veinlets, inner row of spines being second part of mechanism.

PHALLIC COMPLEX: Ectophallus membraneous, except sclerotized cingulum and slightly sclerotized distal part of membrane forming sheath around apical valves of cingulum and apical valves of penis; zygoma moderately short and relatively wide; apodemes moderately long and slender; valves of cingulum present. Endophallus strongly sclerotized; sclerites of penis divided on basal and apical valves, which are rather distant and connected by membrane of endophallic sac; basal valves of penis moderately large, moderately widened and slightly diverging sideways at proximal ends; gonopore processes present; apical valves of penis, almost straight at apices with complicated structures, completely covered with sheath. Epiphallus bridge-shaped; bridge rather wide, sometimes completely divided in middle; ancorae small, angular; lophi narrow, lobiform; lateral plates moderately large; anterior projection small or indistinct.

Spermatheca: Consists of ampoule-like, downcurved main reservoir, with short diverticulum and with very long, narrow spermathecal duct.
Karyotype: 2n♂ = 23.
Type genus: *Hemiacris* Walker, 1870.

The subfamily contains numerous genera.

Distribution: Tropical zone of all World.

Subfamily *Hemiacridinae* was erected by Dirsh in 1956 on the basis of the structure of the phallic complex mostly, divided sclerites of penis and the specialized sound-producing mechanism. However, as further studies revealed both characters are not exclusively to this subfamily. Some Madagascar genera possess a flexure between basal and apical valves of penis and possess also (but reduced) sound-producing mechanism characteristic for *Hemiacridinae*. (Dirsh & Descamps, 1968). In other, particularly from Orental Regions, genera the whole phallic complex is deviating from the typical structure to such an extent that it is difficult to consider them as belonging to the same group (particularly *Tarbaleus* Brunner, 1898). It is no doubt that the subfamily represent heterogeneous complex of the groups, some of them probably unrelated and some of them may be remotely related.

Subfamily is of an ancient origin and mostly of arboreous habitat. Interrelation of the whole subfamily and its groups with other subfamilies can be possible after revising whole this heterogeneous group, which is now under study by Miss J. B. Mason.

Some urgently necessary preliminary division of the subfamily was done in this work.

Subfamily

Spathosterninae

(Fig. 36)

Diagnosis: Body of small size, subcylindrical. Integument smooth. Head conical or subconical; face, in profile, straight or slightly excurved; fastigium of vertex short, at

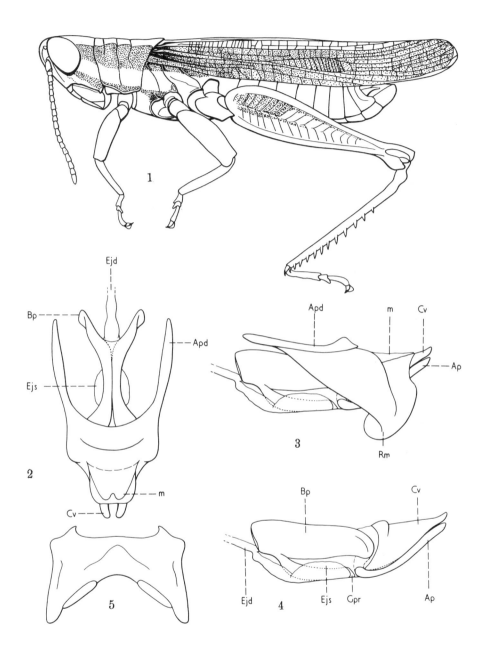

Figure 36.

1, *Spathosternum nigrotaeniatum* (Stal, 1876). 2-5, *Spathosternum pygmaeum* Karsch, 1893, phallic complex. 2, phallic complex, membrane and epiphallus removed, dorsal view. 3, the same, lateral view. 4, endophallus, lateral view. 5, epiphallus, dorsal view.

apex rounded; fastigial foveolae absent. Antennae filiform. Dorsum of pronotum flattened; median and lateral carinae present. Prosternal process present, spathulate. Mesosternal interspace mostly open. Tympanum present. Tegmina and wings fully developed, shortened or lobiform; venation of tegmen in macropterous and brachypterous species specialized. Hind femora moderately slender, with lower basal lobe shorter than upper one. Lobes of hind knee rounded or obtuse angular. External apical spine of hind tibia present. Male cercus narrow conical; supra-anal plate angular; subgenital plate short, subconical. Ovipositor short, valves curved at apices.

Sound-producing mechanism of tegmino-tibial type represented by row parallel thickened, transverse veinlets in radial area of tegmen which are rubbed by the spines in internal side of hind tibia.

PHALLIC COMPLEX: Ectophallus membraneous, except large, strongly sclerotized cingulum; zygoma large, apodemes relatively short and slender, rami large; valves of cingulum present. Endophallus strongly sclerotized; sclerites of penis divided on basal and apical, fully disconnected valves; basal valves moderately large, at proximal ends diverging sideways; gonopore processes present; apical valves relatively long, slightly upcurved. Epiphallus bridge-shaped; bridge wide; ancorae small, angular; lophi lobiform, wide; lateral plates moderately large.

Spermatheca: Main reservoir short and moderately widened; one relatively large and S-curved diverticulum present.
Karyotype: Not known.
Type genus: *Spathosternum* Krauss, 1877.

Besides the type genus, five more genera of this subfamily are known.

Distribution: Ethiopian, Oriental Regions; Australia.

This subfamily was considered as part of the subfamily *Catantopinae*. In 1956 Johnston considered it as a group *Leptacres* which was containing most of the present *Hemiacridinae* and *Tropidopolinae*. Rehn, 1957 erected tribe *Spathosternini*, which here is regarded as a subfamily, mainly on the basis of the phallic complex.

Spathosterninae are related to *Hemiacridinae* by the presence of the common sound-producing mechanism and by the structure of penis' sclerites fully divided on basal and apical valves. Probably the subfamily was separated from *Hemiacridinae* rather late and occupied herbaceous and gramminaceous ecological niche, while *Hemiacridinae* remains mostly arboricolous.

Subfamily

Leptacrinae

(Fig. 37)

Diagnosis: Body from very large to medium, elongated, narrow cylindrical. Head mostly strongly elongated, narrow conical, opisthognathous; face, in profile, mostly straight or slightly incurved or excurved; frontal ridge low, shallowly sulcate or flat; fastigium of vertex from far to moderately protruding forwards, apex angular; fastigial foveolae absent. Antennae mostly ensiform or thick filiform. Dorsum of pronotum cylindrical or flattened; median carina linear, lateral carinae absent. Prosternal process present. Mesosternal interspace closed. Tympanum present. Tegmina and wings fully developed or shortened; reticulation of tegmen moderately sparse; radial area of tegmen with row parallel, thickened, transverse veinlets (stridulatory specialization).

Hind femora slender; lower basal lobe shorter than upper one; lobes of hind knee mostly elongated. External apical spine of hind tibia present. Male cercus of various shape; supra-anal plate angular; subgenital plate from elongated ensiform to comparatively short, acutely conical. Ovipositor short, with curved valves.

Sound-producing mechanism consists of stridulatory veinlets in radial area of tegmen and stridulation possibly achieved by rubbing of inner row of spines of hind tibia against veinlets.

PHALLIC COMPLEX: Ectophallus membraneous except sclerotized cingulum, and slightly sclerotized distal part forming capsule-like covering and sheath for endophallus; zygoma usually short and narrow; apodemes slender and relatively long; valves of cingulum present. Endophallus strongly sclerotized; sclerites of penis divided on basal and apical valves connected only by membrane of endophallic sac; basal valves of penis relatively large, strongly widened and diverging sideways at proximal

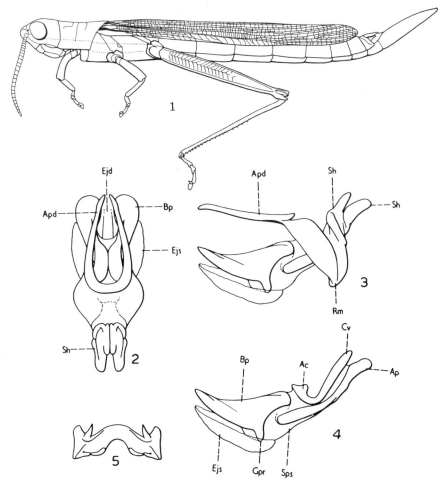

Figure 37.

1, *Leptacris violacea* (Karny, 1907). Male. 2-5, phallic complex of *Leptacris monteiroi* (I. Bolivar, 1890). 2, phallic complex, dorsal view (membrane and epiphallus removed). 3, the same, lateral view. 4, endophallus, lateral view. 5, epiphallus.

ends; gonopore processes present; apical valves of penis slender, regularly upcurved, completely covered with sheath. Epiphallus bridge-shaped; bridge relatively wide, often vertically divided in middle; ancorae small, angular; lophi narrow lobiform; lateral plates relatively large, with large anterior projections.

Spermatheca: Consists of rather narrow, downcurved main reservoir and long, narrow, curved or twisted diverticulum.

Karyotype: Unknown.

Type genus: *Leptacris* Walker, 1870.

The other genera of this new subfamily besides the type genus are *Mesopsera* I. Bolivar, 1908; *Acanthoxia* I. Bolivar, 1906; *Musimoja* Uvarov, 1953; *Oraistes* Karsch, 1896; *Sudanacris* Uvarov, 1944; *Xenippa* Stal, 1878.

Distribution: Ethiopian Region; Oriental Region; Madagascar.

Leptacrinae were considered hitherto as a part of the subfamily *Hemiacridinae* (Dirsh, 1956) on the basis of main common characters; the divided sclerites of penis and the similar sound-producing mechanism. However, both these characters are found in other unrelated or remotely related subfamilies, in which they probably were developed independently in process of evolution. From *Hemiacridinae* the new subfamily differs in most cases by very close proximity of basal and apical valves of penis, by elongate cylindrical shape of body, ensiform antennae, ensiform or narrow, acutely conical male subgenital plate, and closed mesosternal interspace. By the sum of characters of difference this group is recognised here as a separate subfamily *Leptacrinae*.

Subfamily

Conophyminae

(Fig. 38)

Diagnosis: Body short and sturdy, subcylindrical or slightly fusiform, small or medium size. Head short, subglobular or obtusely conical; face, in profile, vertical or slightly oblique, excurved or sometimes concave; frontal ridge concave with weak or without lateral carinulae; fastigium of vertex short, obtuse angular; fastigial foveolae poorly developed. Antennae short, filiform. Dorsum of pronotum subcylindrical sometimes slightly flattened, median carina present, lateral carinae present or partly obliterated. Prosternal process present. Mesosternal interspace wide open. Tympanum absent or rudimentary. Fully apterous. Hind femora short and sturdy, with lower basal lobe shorter than upper; lobes of hind knee rounded. Hind tibia slightly excurved and slightly widened towards distal end; external apical spine present. Male distal abdominal tergite divided. Male supra-anal plate short obtusely subconical.

Sound-producing mechanism not found.

PHALLIC COMPLEX: Ectophallus of various and complicated form partly sclerotized, with cingulum strongly sclerotized; and fully differentiated; valves of cingulum present. Endophallus strongly sclerotized; basal valves relatively small and narrow; gonopore processes present; apical valves relatively long, slightly and regularly curved upwards. Epiphallus bridge-shaped, with bridge from narrow to very wide; ancorae short, not articulated with bridge; lophi small, from lobiform or upcurved angular, or tooth-like.

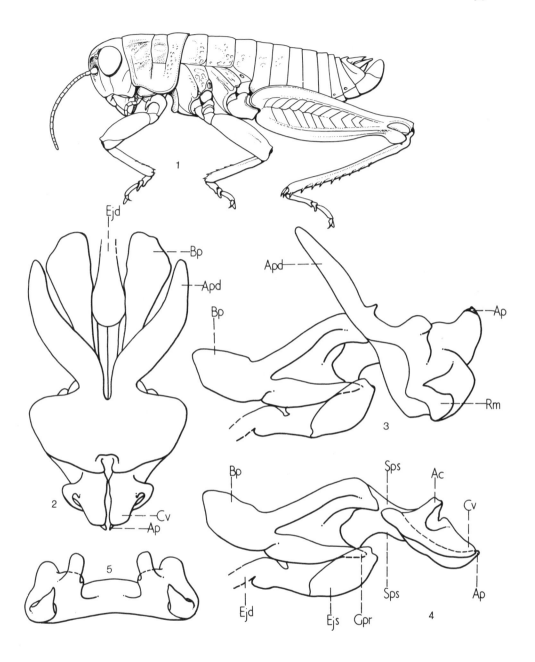

Figure 38.

1, *Conophyma semenovi* Zubovskij, 1898. Male. 2-5, phallic complex. 2, dorsal view (epiphallus and membrane removed). 3, the same, lateral view. 4, endophallus, lateral view. 5, epiphallus.

Spermatheca: Main reservoir ampoule-like, downcurved; one diverticulum present.
Karyotype: Not known.
Type genus: *Conophyma* Zubovskij, 1898.

Only genera *Conophyma* Zubovskij, 1898; *Bienkoa* Mistshenko, 1950; and *Tarbinskia* Mistshenko, 1950 were studied and can be placed into this subfamily with certainty.

Distribution: Mountainous parts of Palearctic Asia.

This group was placed by Jacobson and Bianki, (1904) in the group *Pezotettigini.* In 1952 Mistshenko separated it and raised it to the rank of tribe. Here it is considered as a subfamily.

This subfamily, owing to their nymphal appearance and reduction of the external characters, was considered as a homogeneous group. At present, after studying the phallic complex, it is necessary to divide it into two subfamilies, *Conophyminae* and *Paraconophyminae*, which possess a completely different phallic complex and are not related. Probably, the subfamily after further detailed study of genitalia, will need to be divided again.

The members of subfamily habitate mostly in the mountain region of Central Asia and occur up to 4,200 metres of altitude. Owing to the short period of breeding in high mountainous conditions, the species of the group are most probably neotenic, neoteny being fixed now genetically.

According to the structure of the valves of penis, which are divided and presence of a sheath of penis, this subfamily have certain affinity with the subfamily *Hemiacridinae.* But possibility is not excluded that the similarity in this respect is result of converging evolutionary development.

Subfamily

Chilacrinae

(Fig. 39)

Diagnosis: Body of medium size, dorso-ventrally depressed or subcylindrical. Integument rugose. Head subglobular or subconical; frons, in profile, straight or slightly excurved; frontal ridge flat or shallowly sulcate present; fastigial foveolae absent (sometimes analogous concavities present). Antennae filiform. Dorsum of pronotum flattened; lateral carinae absent. Prosternal process or collar present. Mesosternal interspace wide, short and open. Tympanum present or absent (in apterous species). Tegmina and wings fully developed, strongly shortened or absent. Hind femora mostly rather wide and short, with lower basal lobe as long as upper or slightly shorter; lobes of hind knee obtuse angular or rounded. External apical spine of hind tibia present but very small or absent. Male cercus small, conical; supra-anal plate simple, angular; subgenital plate short, subconical. Ovipositor short, with valves rather slender and slightly curved.

Sound-producing mechanism not found.

Figure 39.   ▶

Chilacris maculipennis Libermann, 1943. Male. 2-5, phallic complex of *Philippiacris rubiosus* Liebermann, 1943. 2, dorsal view (membrane and epiphallus removed). 3, endophallus, dorsal view. 4, the same lateral view. 5, epiphallus. 6, spermatheca of *Chilacris maculipennis*.

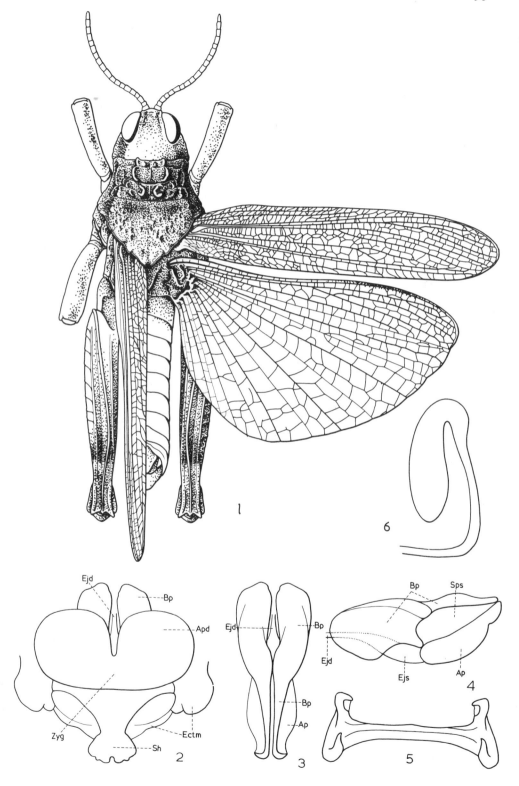

PHALLIC COMPLEX: Ectophallus membraneous, except sclerotized, transverse, shield-like cingulum; no zygoma and no apodemes clearly differentiated; valves of cingulum absent. Endophallus strongly sclerotized; sclerites of penis divided; basal valves of penis wide, robust and form long distal, dorsal projections which extend to point of meeting with apices of apical valves of penis; gonopore processes absent, but anologous projection on ventral distal ends of basal valves exist and at this point are almost touching proximal ends of apical valves of penis; apical valves of penis wide, robust, slightly curved. Spermatophore sac placed between distal dorsal projection of basal valves of penis and apical valves of penis; both, the projections and apical valves fully enclosed in sheath. Epiphallus bridge-shaped; bridge long and narrow; ancorae large, hook-shaped, with hooks incurved; lophi large, tooth-shaped, upcurved; lateral plates small, without anterior projections.

Spermatheca: Represented by large, downcurved, rather wide apical reservoir, without diverticula.
Karyotype: Unknown.
Type genus: *Chilacris* Liebermann, 1943.

Other genera of the subfamily studied by the present author, besides type genus are *Bufonacris* Walker, 1871 and *Philippiacris* Liebermann, 1943. All three genera possess a similar phallic complex and spermatheca.

Distribution: South America.

This subfamily was erected by Liebermann in 1942. He included in it genera *Chilacris* Liebermann, 1943; *Philippiacris* Liebermann, 1943; *Elasmoderus* Saussure, 1888; and *Aucacris* Hebard, 1929. Later genus *Elasmoderus* was excluded by him and genus *Uretacris* Liebermann, 1943 added to the subfamily.
Dirsh, 1961 accepted this arrangement, but added to the subfamily genus *Bufonacris* Walker, 1871.
Eades (1961) excluded genus *Aucacris* from the subfamily, and erected tribe *Aucacrini* transferring it into subfamily (sensu Eades) *Ommexechinae*.
At present work three genera, genitalia of which were studied by the present author, remain as doubtless members of *Chilacrinae*. Genus *Uretacris* was not studied by the present author is omitted.
Peculiar structure of the phallic complex in *Chilacrinae* does not allow for established affinity with any known subfamily of *Acrididae*, except the newly described by Carbonell and Mesa (1972) subfamily *Atacamacrinae*. Both subfamilies possess of the same peculiar structure endophallus.

Subfamily

Atacamacrinae

(Fig. 40)

Diagnosis: Body very small, subcylindrical. Integument rugose. Head subconical, relatively large; face, in profile, almost straight, inclined backwards; frontal ridge shallowly sulcate; fastigium of vertex short and wide, at apex widely rounded; fastigial foveolae absent. Antennae short 8-10 segmented, scape relatively large, apical segment elongated, club-like. Dorsum of pronotum short and wide, without carinae. Prosternum without process or tubercle. Mesosternal interspace wide, open. Tegmina, wings and tympanum absent. Hind femur short and wide; lower basal lobe shorter than upper one. Lobes of hind knee short, obtuse angular. External apical spine of hind tibia absent. Male cercus short, angular; supra-anal plate elongate — angular; subgenital plate short, subconical. Ovipositor moderately short, slender, valves almost straight.

Sound-producing mechanism not detected.

PHALLIC COMPLEX: Ectophallus membraneous and partly sclerotized on ventral surface; cingulum large, zygoma very long, with apodemes sturdy, diverging toward apices, ventral sclerotization relatively large; valves of cingulum absent. Endophallus relatively large, strongly sclerotized; sclerites of penis divided; basal valves large, in proximal part diverging sideways, in distal part forming long narrow projections, apices of which touching with apices of apical valves of penis; and both of them covered with membraneous sheath; gonopore processes absent; distal ends of basal valves are very close to proximal ends of apical valves of penis (possibly they are articulated) , presence of flexure not detected; apical valves of penis narrow, slender, regularly and widely upcurved. Spermatophore sac placed between apical valves of penis and distal projections of basal valves. Epiphallus bridge-shaped; bridge narrow; ancorae small, narrow angular; lophi very small, angularly lobiform; lateral plates large, with long, angular anterior projections.

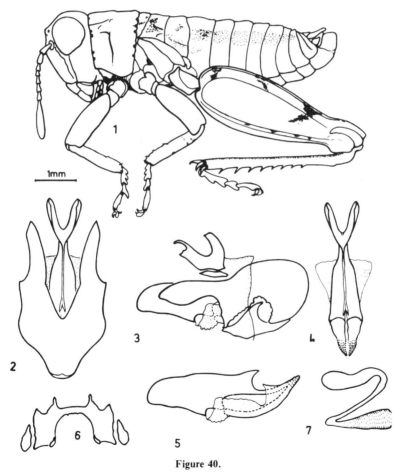

1mm

Figure 40.

1, *Atacamacris diminuta* Carbonell and Mesa, 1972. Male. 2-6, phallic complex. 2, dorsal view (membrane and epiphallus removed). 3, lateral view. 4, endophallus, dorsal view. 5, the same, lateral view. 6, epiphallus. 7, spermatheca. (After Carbonell and Mesa, 1972).

Spermatheca: Simple, tube-like, but widened in apical part.
Karyotype: 2n ♂ = 20.
Type genus: *Atacamacris* Carbonell & Mesa, 1972.
Only the type genus of this subfamily is known.

Distribution: Argentinian Andes, 3,700 metres altitude.

This new subfamily was erected by Carbonell and Mesa in 1972. It cannot be properly fitted into any previously known subfamily of *Acrididae*, but phallic structure, particularly endophallus is similar to that of the subfamily *Chilacrinae* and possibility of the affinity between these two subfamilies cannot be overlooked.

Carbonell and Mesa rightly remarked that the described type genus is probably neotenic. The short season of the life cycle on the altitude 3,700 metres and the nymphal appearance of the insect strongly suggest such possibility.

The representative of this subfamily after *Illapelinae* is a second smallest Acridid, male being 5.8-7.1 and female 9.2-10.6mm.

Subfamily

Leptysminae

(Fig. 41)

Diagnosis: Body elongated, cylindrical, straw-like, of medium size. Head elongate, acutely conical; face, in profile, straight; frontal ridge partly or fully shallowly sulcate; fastigium of vertex elongate angular; fastigial foveolae absent. Antennae ensiform or narrow ensiform. Dorsum of pronotum cylindrical or subcylindrical. Lateral carinae absent. Prosternal process present. Mesosternal interspace closed or semiclosed. Tympanum present. Tegmina and wings fully developed, narrow, tegmen's venation; not specialized; intercalary vein of medial area absent; reticulation sparse. Hind femur narrow, with lower basal lobe shorter than upper; upper lobe of hind knee rounded, lower lobe angular or acute; hind tibia widening toward distal end, external apical spine absent. Male supra-anal plate angular, cercus upcurved, subgenital plate subconical.

Sound-producing mechanism not detected.

PHALLIC COMPLEX: Ectophallus membraneous, except differentiated sclerotized cingulum; valves of cingulum absent. Endophallus strongly sclerotized; penis' sclerites divided on basal and apical valves; basal valves narrow, diverging sideways at proximal ends; gonopore process present; apical valves of penis relatively long and slender, slightly upcurved at apices. Epiphallus bridge-shaped; bridge wide or very wide; ancorae small or rudimentary; lophi relatively large placed at ends, or in middle of bridge.

Spermatheca: With long narrow, downcurved main reservoir, and with one long diverticulum.
Karyotype: Unknown.
Type genus: *Leptysma* Stal, 1873.

This subfamily contains a few genera — *Leptysma* Stal, 1873; *Leptysmina* Giglio-Tos, 1894; *Cylindrotettix* Bruner, 1906; *Stenacris* Walker, 1870; *Inusia* Giglio-Tos, 1897; *Haroldgrantia* Carbonell, 1967.

Distribution: Nearctic and Neotropical Regions.

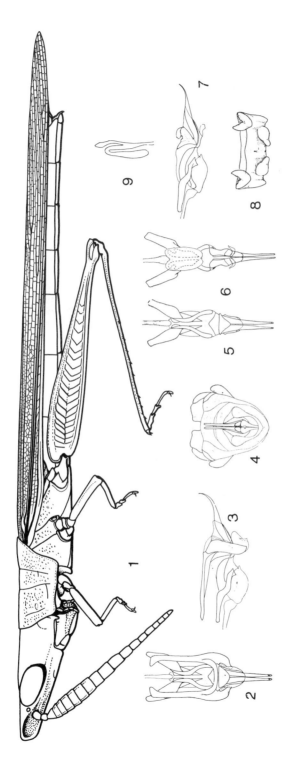

Figure 41.

1, *Leptisma dorsalis* Burmeister, 1838. 2-7, phallic complex of *Haroldgrantia lignosa* Carbonell, Ronderos & Mesa, 1967. 2, phallic complex, dorsal view (membrane and epiphallus removed). 3, the same, lateral view. 4, the same, posterior view. 5, endophallus, dorsal view. 6, the same, ventral view. 7, the same, lateral view. 8, epiphallus. 9, spermatheca. (After Carbonnel, Ronderos and Mesa, 1967).

Leptisminae as a group (Division 10) was first mentioned by Stal, 1877. In the same paper he mentioned also as a group (Division 9) *Opomala* (*Opshomala* Serville, 1831) and *Copiocera* Burmeister, 1838. Since then the later authors treated both groups as one group or tribe *Opomali, Opsomali* or *Opshomali*. Later authors incorporated into this group the Old World genera *Tropidopola* Stal, 1873 and *Leptacris* Walker, 1870, which shared the same elongated shape of body and the similar shape of head.

Old World genera, which were placed in this group, was finally separated from it by Uvarov, 1922, 1926. The New World genera were divided into two groups by Rehn and Eades in 1961. They divided them into groups of genera considered here as subfamilies *Leptisminae* and *Opshomalinae*. Both subfamilies being similar in external appearance, differ greatly in the structure of internal genitalia.

We still do not know enough about this subfamily but it can be suggested that they are branched from the *Acridoidea* stock rather early. The absence of the flexure and separated basal and apical valves of penis strongly indicate to this possibility.

Subfamily

Lithidiinae

(Fig. 42)

Diagnosis: Body small to medium size, with strong sexual dimorphism, males being much smaller than females, subcylindrical or fusiform. Head subglobular; face, in profile, straight or slightly incurved; frontal ridge shallowly sulcate or obliterated in lower part; fastigium of vertex short and wide, at apex widely rounded or with irregularly serrated edge; fastigial foveolae absent. Antennae filiform or slightly widened in apical part. Dorsum of pronotum short, wide, depressed dorso-ventrally; carinae poorly developed or absent. Prosternal process collar-like or absent. Mesosternal interspace wide, open, fused or hardly separated from metasternal interspace. Tympanum absent. Apterous. Hind femur short, wide and inflated, with lower basal lobe much shorter than upper one; lobes of hind knee angular, obtuse at apices. External apical spine of hind tibia absent or present. Males cercus short conical; supra-anal plate angular; subgenital plate short, subconical. Ovipositor short, with relatively slender valves, acute and slightly curved at apices.

PHALLIC COMPLEX: Ectophallus membraneous except sclerotized cingulum; apodeme of cingulum long and narrow, forming elongated U-shaped structure; rami very narrow; valves of cingulum absent. Endophallus moderately strongly sclerotized; basal valves of penis relatively robust, diversing towards proximal end, gonopore processes absent; disconnected with apical valves, which are very narrow, weak and fully enclosed in rather large sheath. Spermatophore sac in dorsal position. Epiphallus bridge-shaped; bridge short and narrow; ancorae short, angular; lophi small, lobiform; lateral plates well developed, with anterior projections hardly noticeable.

Spermatheca have not been studied.
Karyotype: Unknown.
Type genus: *Lithidium* Uvarov, 1925.

Besides the type genus three more genera of this subfamily are known.

Distribution: South and S. West Africa.

This subfamily was erected by Dirsh, 1961 mostly on the basis of the structure of phallic complex, which is rather peculiar and on the basis of external characters which are common with the characters of other subfamilies of *Acrididae* but do not occur in such combination as in *Lithidiinae*.

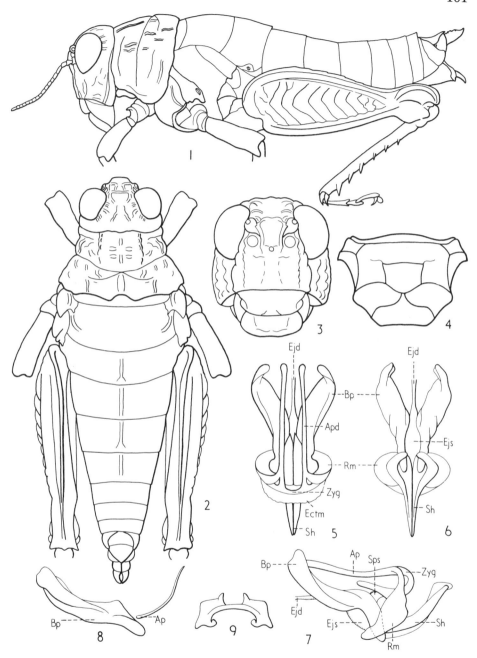

Figure 42.

1-4. *Lithidium bushmanicum* Dirsh, 1956. 1. Female, in profile. 2. The same, from above. 3.
Face. 4. Meso- and metasternum. 5-9. Phallic complex of *Lithidium pusillum* Uvarov, 1925. 5.
From above. 6. From below. 7. In profile. 8. Penis, in profile. 9. Epiphallus.

The interrelation of *Lithidiinae* with other subfamilies is not very clear. Dorsal position of the spermatophore sac and absence of gonopore processes and valves of cingulum may indicate their affinity with *Lentulidae*, from the other hand the divided valves of penis is character which connect them with *Hemiacridinae*.

Family
Catantopidae

Diagnosis: Body of various size and shape. Head of various shape. Fastigial furrow absent; fastigial foveolae absent. Antennae of various shape. Dorsum of pronotum of various form; median carina present or absent, lateral carinae mostly absent. Prosternal process or tubercle present. Mesosternal interspace open or closed. Tympanum mostly present or absent. Tegmina and wings fully developed, shortened, lobiform, vestigial or absent. Lower basal lobe of hind femur shorter or of same length as upper one.

Sound-producing mechanism mostly of wing-tegminal type, abdomino-tibial or tegmino-tibial type.

PHALLIC COMPLEX: Ectophallus partly membraneous and partly sclerotized; cingulum present, well differentiated; valves of cingulum present or absent. Endophallus with bi-sclerited penis, divided on basal and apical valves connected by flexure. Gonopore processes present or absent. Spermatophore sac in middle position. Epiphallus of various shape; ancorae not articulated, often absent; lophi present. Oval sclerites present.

Spermatheca of various shape.
Type genus: *Catantops* Schaum, 1853.

Distribution: Whole World.

List of subfamilies

1. *Anamesacrinae*
2. *Apoboleinae*
3. *Aucacrinae*
4. *Calliptaminae*
5. *Catantopinae*
6. *Coptacrinae*
7. *Cyrtacanthacrinae*
8. *Dericorythinae*
9. *Diexinae*
10. *Egnatiinae*
11. *Euryphyminae*
12. *Eyprepocneminae*
13. *Galideinae*
14. *Illapelinae*
15. *Opshomalinae*
16. *Oxyinae*
17. *Paraconophyminae*
18. *Pargainae*
19. *Podisminae*
20. *Romaleinae*

21. *Shelfordinae*
22. *Teratodinae*
23. *Tropidopolinae*

Key to subfamilies

1 (12)   Lower basal lobe of hind femur as long as upper one, or insignificantly longer or shorter.

2 (11)   Tegmina and wings fully developed, shortened or lobiform, rarely absent. Tympanum normally present.

3 (6)   Sound-producing mechanism of wing-tegminal type present.

4 (5)   External apical spine of hind tibia present.
Romaleinae

5 (4)   External apical spine of hind tibia absent.
Teratodinae

6 (3)   Sound-producing mechanism absent, or not detected.

7 (10)   Valves of cingulum present. Dorsum of pronotum subcylindrical, crested, or tuberculate.

8 (9)   Bridge of epiphallus semi-divided. Tegmina lobiform. Dorsum of pronotum subcylindrical.
Diexinae

9 (8)   Bridge of epiphallus solid. Tegmina and wings fully developed or shortened. Dorsum of pronotum crested or tuberculate.
Dericorythinae

10 (7)   Valves of cingulum absent. Dorsum of pronotum flattened.
Aucacrinae

11 (2)   Tegmina and wings and tympanum absent.
Anamesacrinae

12 (1)   Lower basal lobe of hind femur shorter than upper one.

13 (16)   Sound-producing mechanism present.

14 (15)   Sound-producing mechanism of abdomino-femoral type. Mesosternal interspace shortened.
Egnatiinae

15 (14)   Sound-producing mechanism of tegmino-tibial type. Mesosternal interspace of normal shape.
Apoboleinae

16 (13)   Sound-producing mechanism absent or not detected.

17 (40)   Tegmina and wings fully developed, shortened, vestigial rarely absent. Tympanum present or absent.

18 (23)   Body strongly elongated stick-like.

19 (22)   Mesosternal interspace closed.

20 (21)   Tympanum present. Epiphallus with lobiform lophi.
Tropidopolinae

21 (20) Tympanum absent. Epiphallus with hook-shaped, incurved lophi.

Galideinae

22 (19) Mesosternal interspace open.

Opshomalinae

23 (18) Body short, subcylindrical, not stick-like.

24 (25) Lower outer lobe of hind knee at apex spine-like. Epiphallus in middle divided.

Oxyinae

25 (24) Lower outer lobe of hind knee of various shape but not spined. Epiphallus in middle solid or divided.

26 (27) Posterior margin of male last abdominal tergite (in majority of genera) with furcula. Supra-anal plate with attenuate or trilobate apex. Subgenital plate with transverse fold. Epiphallus divided or with tendency to division.

Coptacrinae

27 (26) Posterior margin of male last abdominal tergite without furcula. Supra-anal plate of various shape. Subgenital plate without transverse fold. Epiphallus of various shape.

28 (29) Male cerci pincer-like adapted for grasping. Epiphallus discoidal.

Calliptaminae

29 (28) Male cerci of various shape, but not pincer-like. Epiphallus not discoidal.

30 (31) Male cercus with large basal articulation. Posterior margin of last abdominal tergite in male strongly sclerotized. Epiphallus in middle divided.

Euryphyminae

31 (30) Male cercus with small basal articulation. Posterior margin of last abdominal tergite with normal sclerotization. Epiphallus of various shape.

32 (33) Male cercus with lobiform, rounded or subacute, downcurved apex. Epiphallus mostly weakly sclerotized, and mostly with wide bridge.

Eyprepocneminae

33 (32) Male cercus of different and various shape. Epiphallus strongly sclerotized with ancorae if present not articulated.

34 (35) Mesosternal lobes rectangular.

Cyrtacanthacrinae

35 (34) Mesosternal lobes rounded or obtuse angular.

36 (39) Antennae mostly filiform or sometimes narrow ensiform not specialized.

37 (38) Antennae filiform. External apical spine of hind tibia absent.

Podisminae

38 (37) Antennae filiform or slightly ensiform. External apical spine of hind tibia present or absent.

Catantopinae

39 (36) Antennae specialized, ensiform and differentiated, with flagellum mostly divided on basal, medial and apical parts.

Pargainae

40 (17) Tegmina and wings strongly reduced, lobiform or absent. Tympanum present or absent.

41 (42) Tegmina and wings strongly reduced. Tympanum present.

Paraconophyminae

42 (41) Tegmina and wings absent. Tympanum absent.

43 (44) Larger, bifurcate appendage between split of two last abdominal tergites, in male, present.

Shelforditinae

44 (43) Very small (5.5mm). Bifurcate appendage absent.

Illapelinae

Subfamily

Romaleinae

(Fig. 43)

Diagnosis: Body from large to medium size. Head of variable shape; face, in profile, mostly incurved; frontal ridge sulcate; fastigium of vertex at apex angular, sometimes strongly protruding forwards; fastigial foveólae absent. Antennae filiform or slightly ensiform. Dorsum of pronotum of variable shape sometimes with high of various shape carinae. Prosternal process present. Mesosternal interspace open. Tympanum present, in apterous species absent. Tegmina and wings fully developed, shortened, rarely absent; membrane mostly parchment-like; reticulation dense; hind wing specialized for sound-producing mechanism. Hind femora from stout to slender, with lower basal lobe as long as upper one; lobes of hind knee rounded or angular; external apical spine of hind tibia present. Male cercus mostly conical. Supra-anal plate angular. Subgenital plate subconical. Ovipositor short, with valves curved.

Sound-producing mechanism represented by first vannal area of hind wing being narrow, convex, forming tube-like fold, when wing is folded, this area contains arched, parallel, finely serrated transverse veinlets, sometimes adjoining longitudinal veins are also serrated. Cubital, second vannal, and sometimes medial area are widened and covered with thickened, regular, transverse veinlets. Sound produced by rubbing the serrated veinlets against sharp veins of inner surface of tegmen.

PHALLIC COMPLEX: Ectophallus partly membraneous, partly with sclerotized dorsal side and strongly sclerotized fully differentiated cingulum; apodemes of cingulum short, robust. Valves of cingulum mostly absent. Endophallus strongly sclerotized; basal valves of penis wide, expanding upwards and moderately spread sideways; gonopore processes present. Apical valves of penis relatively slender, slightly upcurved. Flexure long and rather robust. Epiphallus bridge-shaped, with short, moderately wide bridge; ancorae small, angular, at apices obtuse or acute; lophi lobiform or tooth-like; lateral plates moderately developed, anterior projections small or hardly developed.

Spermatheca: Consists of single downcurved and twisted main reservoir and in place of connection with spermathecal duct small protruding inflation which can be considered as rudimentary diverticulum.

Karyotype: $2n\ \male = 23$.

Type genus: *Romalea* Serville, 1831.

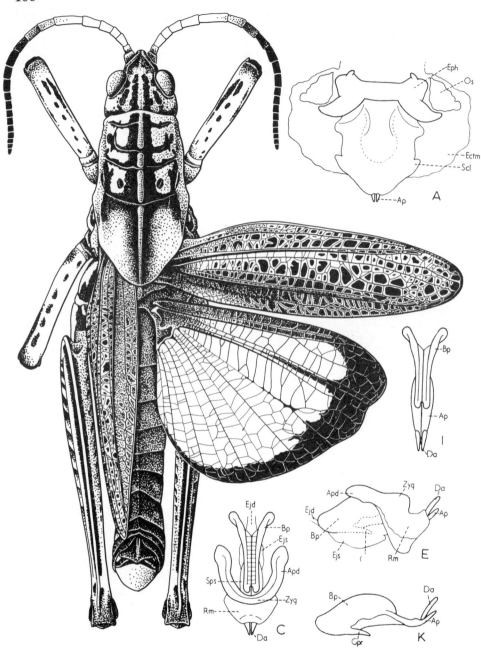

Figure 43.

Romalea microptera (Beauvais, 1805). Male. *A*, whole phallic complex, dorsal view. *C*, phallic complex, dorsal view (membrane and epiphallus removed). *E*, the same, lateral view. *I*, endophallus, dorsal view. *K*, the same, lateral view.

This subfamily contains about forty well defined genera.

Distribution: North and South America.

Subfamily *Romaleinae* was erected by Roberts in 1941 on the basis of phallic complex.

Rehn and Grant in 1959 divided the subfamily into 17 tribes comprising 39 genera. This division was not recognized by most acridologists owing to the reason that six tribes from the proposed seventeen were monogeneric and five of them bigeneric, which made the tribes value as the taxa very doubtful (Uvarov & Dirsh, 1961).

Dirsh, 1956 redefined the subfamily adding a new character — a peculiar sound-producing mechanism and included into subfamily group *Teratodini* of Old World.

Despite the Old World genera possess the same sound-producing mechanism as the New World *Romaleine*, most acridologists disagree with such arrangement and prefer to consider the Old World genera as a tribe *Teratodini*.

The difference between *Romaleinae* and *Teratodini* is that the former possess the external apical spine of hind tibia, while in the latter it is absent. The valves of cingulum in majority of *Romaleinae* are absent, while in *Teratodinae* they are present. Of course the hyatus in geographical distribution is also one of the reasons for separating them.

Romaleinae however are closely related to *Teratodinae* which are treated here as a separate subfamily.

Interrelation of *Romaleinae* with other subfamilies of *Acrididae* is not clear, but probably they have affinity with subfamily *Cyrtacanthacrinae*.

Subfamily

Teratodinae

(Fig. 44)

Diagnosis: Body from large to medium size, subcylindrical or laterally compressed. Head mostly subglobular; face, in profile, straight, slightly inclined backwards; frontal ridge shallowly sulcate or flat; fastigium of vertex short widely angular; fastigial foveolae indistinct. Antennae short, filiform. Dorsum of pronotum crest-like, tectiform or flattened, median carina present or absent, lateral carinae absent. Prosternal process present. Mesosternal interspace open. Tympanum present, sometimes rudimentary. Tegmina and wings fully developed, shortened or strongly reduced, reticulation rather sparse; in winged species hind wing specialized forming sound-producing mechanism. Hind femora sturdy and wide, with lower basal lobe slightly shorter than upper; lobes of hind knee rounded at apices. External apical spine of hind tibia absent. Male cercus short, conical; supra-anal plate short, widely angular; subgenital plate short subconical. Ovipositor short, with valves acute and curved at apices.

Sound-producing mechanism as in *Romaleinae*.

PHALLIC COMPLEX: Ectophallus membraneous except strongly sclerotized cingulum, with very long zygoma; valves of cingulum present. Endophallus strongly sclerotized; basal valves of penis moderately wide, at proximal ends protruding sideways; gonopore processes present; apical valves of penis large and widened towards apex. Flexure relatively short, in middle narrow. Epiphallus bridge-shaped; bridge relatively wide; ancorae small, angular; lophi lobiform relatively wide; lateral plates small, anterior projections well developed.

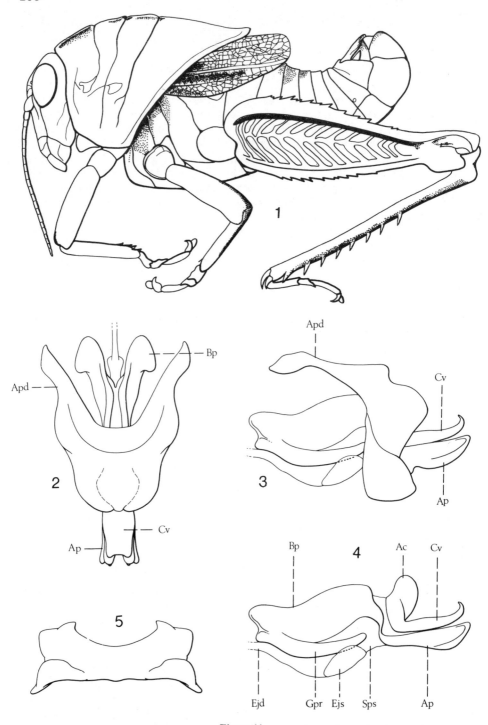

Figure 44.

1, *Acrostegaster glaber* Karsch, 1896. Male. 2-5, phallic complex of *Teratodes monticollis* (Gray, 1832). 2, phallic complex, dorsal view (membrane and epiphallus removed). 3, the same, lateral view. 4, endophallus, lateral view. 5, epiphallus.

Spermatheca: With relatively narrow, downcurved main reservoir and two or several twisted diverticula.
Karyotype: Unknown.
Type genus: *Teratodes* Brulle, 1835.

This subfamily contains besides the type genus three African and five Asian genera.

Distribution: S.W. Asia and N.E. Africa.

Teratodinae were recognised as a tribe of *Catantopinae* by Mistshenko in 1952. Dirsh in 1956 incorporated it in the subfamily *Romaleinae*. Since then the attitude of acridologists were divided, some of them considered *Teratodinae* as a tribe of *Catantopinae* and some as a part of the subfamily *Romaleinae*. At present state of our knowledge of the group, it is preferable to consider it as a separate subfamily on the basis of their genital organs and other characters mentioned in diagnosis. The common with *Romaleinae* character — the similar sound-producing mechanism may be a result of convergence in evolution of this group or a result of earlier divergence of both subfamilies from the common stock.
The subfamily, however, is related to *Romaleinae* and *Cyrtacanthacrinae*.

Subfamily

Diexinae

(Fig. 45)

Diagnosis: Small or medium size. Body elongated cylindrical or fusiform. Head narrow conical or subconical; face, in profile, straight or incurved; fastigium of vertex short, at apex angular or rounded; fastigial foveolae absent. Antennae short, filiform. Dorsum of pronotum subcylindrical, widening towards posterior end, lateral carinae absent. Prosternal process short or substituted in front by collar-like elevation. Mesosternal interspace short and wide. Tympanum present. Tegmina and wings strongly shortened, lateral; membrane and reticulation coarse; venation reduced. Hind femora narrow, basal lower and upper lobes of almost equal length; upper lobe of hind knee rounded, lower lobe angular. Hind tibia slightly curved and widening towards distal end, external apical spine present. Male cercus short, conical; supra-anal plate angular; subgenital plate subconical. Ovipositor short, rather sturdy with valves curved at apices.
Sound-producing mechanism not detected.

PHALLIC COMPLEX: Ectophallus membraneous except strongly sclerotized cingulum; zygoma long and wide; apodemes relatively short, incurved at proximal ends; rami wide; valves of cingulum present. Endophallus strongly sclerotized; basal valves of penis relatively narrow, at proximal ends curved sideways; gonopore processes present, relatively very long; apical valves of penis long and from above comparatively very wide; flexure gradually merging with basal and apical valves. Epiphallus bridge-shaped, bridge wide, semi-divided in middle; ancorae short, stout, at apices obtuse not articulated with bridge; lophi lobiform, transverse, monolobate; lateral plates comparatively narrow.

Spermatheca: With main reservoir of oval form downcurved, with one diverticulum.
Karyotype: Unknown.
Type genus: *Diexis* Zubovskij, 1899.

At present only two genera of this subfamily are known — the type genus and *Bufonacridella* Adelung, 1910.

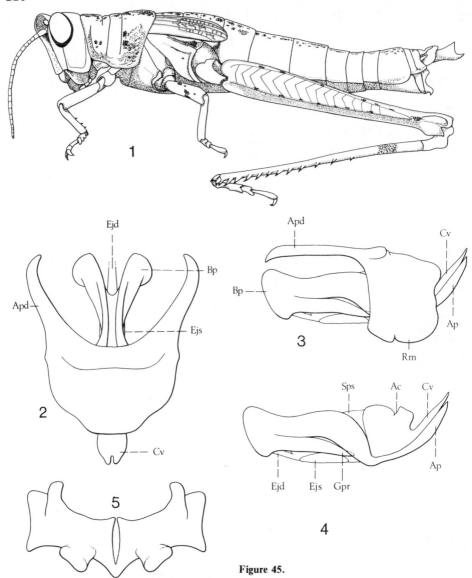

Figure 45.

Diexis varentzovi Zubovskij, 1899. 1, female. 2-5, phallic complex. 2, phallic complex, dorsal view (membrane and epiphallus removed). 3, the same, lateral view. 4, endophallus, lateral view. 5, epiphallus.

Distribution: Palaearctic Region (Republic of U.S.S.R.: Kazakstan, Turkmenistan, Uzbekistan and Tadzhikistan, Iran; Afghanistan).

This subfamily was recognised as a tribe of the subfamily *Catantopinae* by Mistshenko in 1947. He based his definition on the external characters. However, the structure of the phallic complex, with very wide poorly differentiated flexure make it necessary to elevate the tribe to subfamily rank.

The interrelation of *Diexinae* with other subfamilies is rather obscure and, at present, no subfamily even remotely related to this subfamily can be found. Probably, they are an ancient relic of a past fauna.

Subfamily

Dericorythinae

(Fig. 46)

Diagnosis: Body from large to medium size. Head subglobular; face, in profile, straight; frontal ridge flat or slightly sulcate; fastigium of vertex short; fastigial foveolae absent. Antennae filiform; or widening towards apex. Dorsum of pronotum

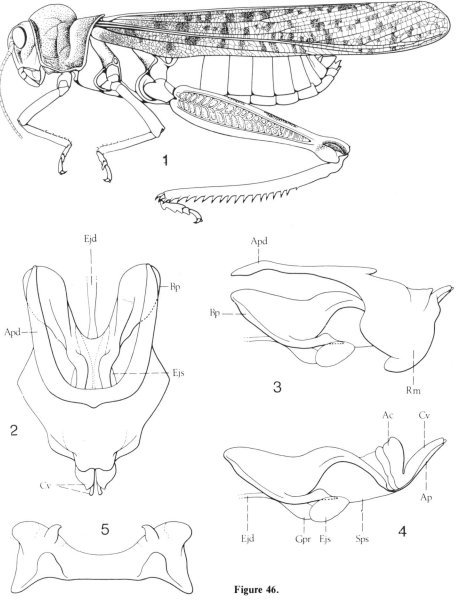

Figure 46.

1, *Dericorys albidula* Serville, 1838. Female. 2-5, phallic complex. 2, phallic complex, dorsal view (membrane and epiphallus removed). 3, the same, lateral view. 4, endophallus, lateral view. 5, epiphallus.

in prozona crest-like, or with tubercle, lateral carinae absent. Prosternal process present. Mesosternal interspace open. Tympanum present. Tegmina and wings fully developed or shortened; reticulation of tegmen dense. Venation mostly not specialized. Hind femora moderately slender, with lower basal lobe slightly longer or as long as upper one. Lobes of hind knee rounded. Hind tibia slightly curved, with external apical spine present. Male cercus short, conical. Supra-anal plate simple, elongate angular, subgenital plate short subconical. Ovipositor short, with curved valves, acute at apices; sometimes valves red or blue coloured.

Semblance of sound-producing mechanism consisting of inflated part of subcostal area of hind wing and convex veins of inner surface of tegmen, occur in some species.

PHALLIC COMPLEX: Ectophallus membraneous except sclerotized cingulum; apodemes of cingulum long, narrow; cingular valves relatively robust. Endophallus strongly sclerotized; basal valves of penis large, strongly widened and spread sideways at proximal ends; gonopore processes present. Apical valves of penis slender, narrow, upcurved; flexure long and relatively robust. Epiphallus bridge-shaped, bridge comparatively long and moderately narrow; ancorae small, incurved, with apices acute; lophi lobiform, upcurved, lateral plates small, anterior projections relatively long.

Spermatheca: Spermathecal duct in apical part widened; main reservoir large, relatively narrow, twisted and downcurved; one diverticulum present, short and slightly curved.
Karyotype: Unknown.
Type genus: *Dericorys* Serville, 1838.

Only three genera of this subfamily are known — the type genus, *Corystoderes* I. Bolivar, 1936 and *Farsinella* Bey-Bienko, 1948.

Distribution: Northern part of Africa approximately down to 20° N. latitude, and costal zone of S.E. Africa down to 10° N. latitude. Canary Is., S.W. and Central Asia.

This subfamily was erected by Jakobson & Bianki in 1904. Later authors placed it into *Catantopinae*. Dirsh, 1961 restored it as a subfamily.
Subfamily *Dericorythinae* is not closely related to any other subfamily of *Acrididae*. However the structure of the phallic complex with its robust endophallus and wide, long flexure, indicates a remote possibility of affinity with the subfamily *Romaleinae*.
Ecological niche of the subfamily — the desert and semi-desert areas.

Subfamily

Aucacrinae

(Fig. 47)

Diagnosis: Body of medium size, subcylindrical or slightly fusiform. Integument slightly rugose. Head subglobular; face, in profile, straight; frontal ridge shallowly sulcate; fastigium of vertex with faint trace of fastigial furrow; fastigial foveolae absent. Antennae filiform. Dorsum of pronotum flattened, without lateral carinae. Prosternal process present. Mesosternal interspace open, wide. Tympanum present. Tegmina and wings shortened; reticulation of tegmen dense; venation not specialized. Hind femora moderately sturdy; with lower basal lobe of the same length as upper; lobes of hind knee rounded. External apical spine of hind tibia present. Male cercus simple, conical; supra-anal plate angular; subgenital plate short, subconical. Ovipositor short, with valves curved at apices.
Sound-producing mechanism not found.

PHALLIC COMPLEX: Ectophallus in proximal part membraneous, in distal part slightly sclerotized forming sheath covering distal part of endophallus; cingulum strongly sclerotized with elongate, narrow zygoma; apodemes short, curved, slightly diverging backwards; valves of cingulum absent. Endophallus strongly sclerotized; basal valves of penis narrow and sturdy; gonopore processes present, short; flexure very thick and sturdy; apical valves of penis short, wide; upcurved, with subacute apices. Epiphallus almost disc-shaped, with hint on the presence of very wide bridge; ancorae small, angular, at apices acute; lophi large, elongate lobiform or irregular shape; lateral plates large; anterior projections small.

Spermatheca: Main reservoir elongate, curved down and curved in apical part; diverticulum short and relatively wide.

Karyotype: Unknown.

Type genus: *Aucacris* Hebard, 1929.

Besides the type genus, only genus *Neuquenia* Rehn, 1943 can be placed in this subfamily with certainty.

Distribution: South America.

The type genus *Aucacris* first described by Hebard in 1929 was placed by him in subfamily *Cyrtacantacrinae* (=*Cantantopinae*). Uvarov, (1937) transferred it into

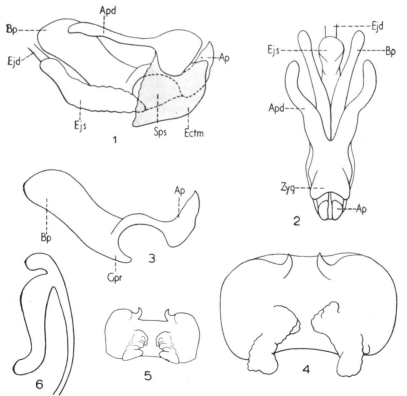

Figure 47.

Aucacris eumera Hebard, 1929 (type): 1, phallic complex, lateral view; greater part of ectophallic membrane and epiphallus removed; 2, the same, from above; 3, penis, lateral view; 4, epiphallus; 5, the same epiphallus at a different angle and smaller magnification; 6, spermatheca. (Fig. 1-4, the same magnification).

group *Batrachotetrigini.* Lieberman in 1942 placed it into his newly erected subfamily *Chilacrinae.* Rehn (1943) considered *Aucacris* as a genus of the subfamily *Cyrtacanthacrinae* together with genera *Cumainocloidus* Brunner, 1913 and *Neuquenia* Rehn, 1943. Dirsh (July, 1961) referred the genus to the Liebermann's subfamily *Chilacrinae* together with genera *Bufonacris* Walker, 1871 *Chilacris* Liebermann, 1943; *Philippiacris* Liebermann 1943, and *Uretacris* Liebermann, 1943. Eades (November, 1961) raised the Rehn's group *Aucacris* to tribal rank *Aucacrini,* and placed the tribe into subfamily (sensu Eades) *Ommexechinae.* The latter point of view is totally unacceptable owing to a great difference in the structure of the phallic complex between *Ommexechidae* and *Aucacrini.*

The present author after studying the genitalia of the type of *Aucacris eumera* Hebard, 1929, and the genitalia of *Philippiacris rubiosus* Liebermann, 1943 came to the conclusion that subfamily *Chilacrinae* (sensu Dirsh 1961) is heterogeneous complex which cannot be placed in the same taxon. Accordingly the genus *Aucacris* and *Neuquinia* on the basis of characteristic structure of the phallic complex are considered here as a separate subfamily.

At present knowledge of *Acridoidea* it is rather difficult to find affinity between *Aucacrinae* and other subfamilies of the family. The endophallus of the *Aucacrinae* however, with its very sturdy flexure, suggests possibility of the affinity with *Romaleinae.*

Subfamily

Anamesacrinae

(Fig. 48)

Diagnosis: Body small, subcylindrical. Integument rugose and tuberculate. Head conical; face, in profile, incurved; frontal ridge sulcate; fastigium of vertex angular, far protruding forwards; fastigial foveolae absent. Antennae short, ensiform or rod-like. Dorsum of pronotum rugose, in prozona with hump or tubercle; median and lateral carinae irregular. Prosternum with collar-like convexity. Mesosternal interspace open. Tympanum absent. Tegmina and wings absent. Hind femora moderately slender, with lower basal lobe as long or slightly longer than upper one; lobes of hind knee rounded or obtuse angular. Hind tibia straight, external apical spine present. Male cercus short, conical. Supra-anal plate elongate-angular. Subgenital plate short, subconical. Ovipositor short, slender, valves curved, at apices acute.

Sound-producing mechanism not found.

PHALLIC COMPLEX: Ectophallus membraneous except sclerotized cingulum; zygoma of cingulum relatively short and wide; apodemes short and narrow; rami wide; valves of cingulum present. Endophallus strongly sclerotized; basal valves of penis sclerites long and relatively narrow, moderately diverging at proximal ends; gonopore processes present; apical valves short and slender; flexure short and narrow. Epiphallus bridge-shaped; bridge short and narrow; ancorae short, angular obtuse at apices; lophi very wide transverse lobiform; lateral plates exceptionally large; anterior projections small.

Spermatheca: Not known.
Karyotype: Unknown.
Type genus: *Anamesacris* Uvarov, 1934.

Besides type genus this subfamily contains genera *Pamphagulus* Uvarov, 1929 and *Bolivaremia* Morales Agacino, 1949.

Distribution: Northern part of Africa, down to Tropic of Cancer.

The subfamily was considered as a part of the subfamily *Dericorythinae* (Dirsh, 1961), but the difference in the structure of the phallic complex convinced the present author that such arrangement is artificial and that the group must be considered as a separate taxon of a subfamily rank.

On existing data concerning the subfamily it is not possible to decide its affinity with other subfamilies of *Acridoidea*.

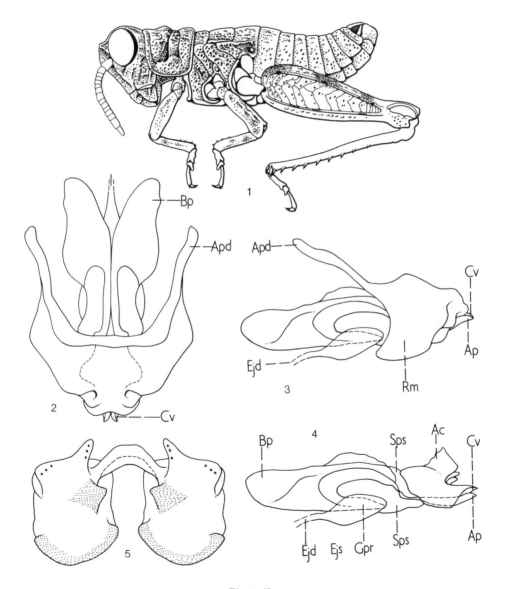

Figure 48.

1, *Anamesacris abajoi* Morales Agacino, 1949. Male. 2-5, phallic complex. 2, phallic complex dorsal view (membrane and epiphallus removed). 3, the same, lateral view. 4, endophallus, lateral view. 5, epiphallus.

Subfamily

Egnatiinae

(Fig. 49)

Diagnosis: Body small, subcylindrical. Integument moderately rugose. Head subconical; face, in profile, straight or slightly excurved; fastigium of vertex short; fastigial foveolae present, sometimes indistinct. Antennae filiform sometimes slightly clavate. Dorsum of pronotum slightly saddle-shaped; median carina present; lateral carinae mostly present. Low prosternal tubercle present. Mesosternal interspace open, very short. Tympanum present. Tegmina and wings fully developed or shortened; venation of tegmen not specialized, weak intercalary vein of medial area present; wing venation with subcostal, median and cubital area expanded or not specialized. Lateral sides of abdomen sometimes with row of transverse ridges. Hind femora moderately slender or widened; lower basal lobe shorter than upper one. Lobes of hind knee rounded or obtuse angular. External apical spine of hind tibia absent. Male cercus short, subconical; supra-anal plate angular, subgenital plate subconical. Ovipositor short, with valves curved at apices.

Sound-producing mechanism represented by transverse ridges on sides of abdominal tergites. In case of specialized venation on hind wing this specialization probably is part of another kind of sound-producing mechanism.

PHALLIC COMPLEX: Ectophallus membraneous, except strongly sclerotized, well differentiated cingulum; zygoma wide; apodemes relatively long, rami narrow; valves of cingulum present. Endophallus strongly sclerotized; penis' sclerites divided on basal and apical valves, connected by sturdy flexure; basal valves of penis very wide, at proximal ends strongly curved sideways; gonopore processes present; apical valves relatively wide, upcurved. Epiphallus bridge-shaped; bridge narrow; ancorae slightly curved, articulated with bridge; lophi large, lobiform; lateral plates moderately large; anterior projections large.

Spermatheca: Unknown.
Karyotype: Unknown.
Type genus: *Egnatius* Stal, 1876.

Besides the type genus, five genera of this subfamily are known.

Distribution: North Africa, S.W. and Central Asia.

The representatives of this group were considered as the members of subfamily *Oedipodinae*. Bey-Bienko in 1951 erected for them a new subfamily — *Egnatiinae*. At this rank the group is considered now.

Peculiar form of furcal sture of mesosternum and presence of original sound-producing mechanism in some of its members, separate this subfamily from other subfamilies of *Acridoidea*.

Slifer, 1939 discovered that they possess a Comstock-Kellog gland, which at that time were supposedly present only in the subfamily *Catantopinae*. This character was connecting *Egnatiinae* with that subfamily. Bryantseva, 1953 also found that the folds and sculpture of internal surface in foregut in *Egnatiinae* are similar to those in *Catantopinae* and different to those in *Oedipodinae*. These characters are linking *Egnatiinae* with *Catantopinae* more than with any other subfamily.

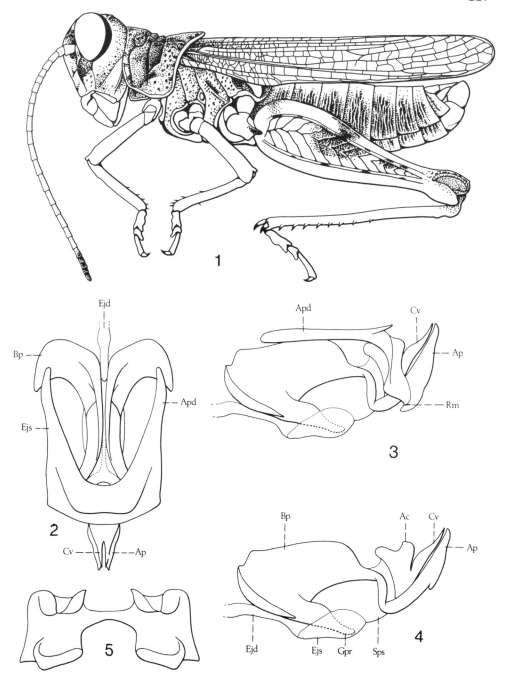

Figure 49.

1, *Egnatius apicalis* Stal, 1876. Male. 2-5, phallic complex of *Paraignatius salavatiani* Popov, 1951. 2, phallic complex, dorsal view (membrane and epiphallus removed). 3, the same, lateral view. 4, endophallus, lateral view. 5, epiphallus.

118
Subfamily

Apoboleinae

(Fig. 50)

Diagnosis: Body subcylindrical, small or medium size. Head short, subconical; face, in profile, oblique with upper part protruding forwards; frontal ridge with or without lateral carinulae, below median ocellus sometimes obliterated; fastigium of vertex short, at apex angular or truncate; fastigial foveolae absent. Antennae filiform. Dorsum of pronotum subcylindrical or slightly flattened; lateral carinae absent or very poorly developed. Prosternal process present. Tegmina and wings from submacropterous to micropterous; venation straight, reticulation rather dense; costal area of tegmen in submacropterous species, widened with row oblique, parallel

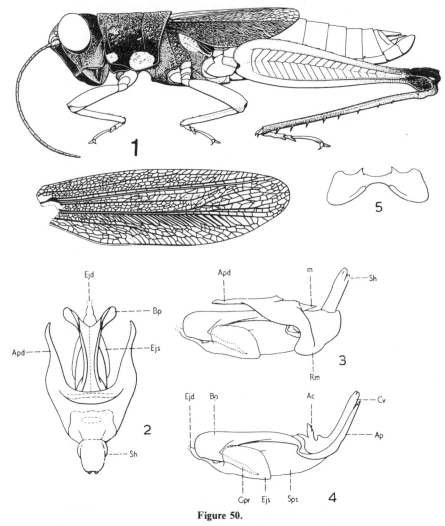

Figure 50.

1, *Apoboleus sudanensis* Dirsh, 1952. Male and male tegmen. 2-5, phallic complex of *Apoboleus affinis* Kevan, 1955. 2, phallic complex, dorsal view (membrane and epiphallus removed). 3, the same, lateral view. 4, endophallus, lateral view. 5, epiphallus.

veinlets; wings shortened, with reduced venation and widely rounded outer margin. Hind femora slender, with lower basal lobe shorter than upper one; lobes of hind knee rounded; hind tibia straight; external apical spine absent. Male supra-anal plate simple, angular; cercus short conical or incurved; subgenital plate short, subconical.

Sound-producing mechanism probably is represented by specialized venation of costal area of tegmina.

PHALLIC COMPLEX: Ectophallus membraneous except strongly sclerotized well differentiated cingulum; zygoma narrow; apodemes relatively short; rami wide; valves of cingulum present. Endophallus strongly sclerotized; basal valves of penis relatively slender at proximal ends slightly curved sideways; gonopore processes present. Apical valves of penis relatively long, upcurved and, together with long valves of cingulum, enclosed in sheath; flexure short. Epiphallus bridge-shaped; bridge narrow; ancorae short, not articulated with bridge; lophi narrow lobiform; lateral plates large and wide.

Spermatheca: Unknown.
Karyotype: Unknown.
Type genus: *Apoboleus* Karsch, 1891.

Distribution: Ethiopian and Oriental Regions, Madagascar.

This subfamily is a group of genera which previously were divided into two groups — *Apobolei* and *Serpusiae*, within the subfamily *Catantopinae*. Basis for separating this subfamily is characteristic venation of tegmina in *Apobolei*, probably owing to micropterism this kind of venation is undetectable in *Serpusiae*.

The structure of the phallic complex however is common to both groups and allow them to be considered as a taxon of subfamily rank.

From the subfamily *Catantopinae*, with which it has affinity, the *Apoboleinae* differs by the venation of tegmina, shape of head, heavily sheathed cingular and apical valves of penis and partly by the shape of epiphallus.

Subfamily

Tropidopolinae

(Fig. 51)

Diagnosis: Body from large to small size, elongated cylindrical or subcylindrical. Integument smooth. Head conical to elongated narrow conical, sometimes subglobular; face, in profile, slightly excurved or straight; frontal ridge flat or shallowly sulcate; fastigium of vertex from short obtuse angular to strongly elongated narrow angular; fastigial foveolae absent or present. Antennae filiform or ensiform. Dorsum of pronotum subcylindrical or flattened, median carina present, linear; lateral carinae present or absent. Prosternal process present. Mesosternal interspace closed. Tympanum present. Tegmina and wings fully developed or shortened; reticulation moderately sparse; venation not specialized. Hind femora slender, with lower basal lobe shorter than upper one; lobes of hind knee rounded or obtuse angular. External apical spine of hind tibia present. Male cercus of variable shape; supra-anal plate from elongated acutely conical to short, subconical, with rounded apex. Ovipositor from short, with curved valves, to relatively long, with straight valves.

Sound-producing mechanism not found.

PHALLIC COMPLEX: Ectophallus membraneous, except sclerotized cingulum — and sometimes with slightly sclerotized apical part; zygoma of cingulum of variable shape;

120

apodemes relatively short; valves of cingulum of various shape present, rami usually large. Endophallus strongly sclerotized; basal valves of penis wide, in proximal part widely spreading sideways — connected with apical valves by short or relatively long flexure, which sometimes have tendency to disappear; gonopore processes present. Apical valves of penis straight, upcurved or downcurved, usually covered with sheath. Epiphallus bridge-shaped; bridge relatively wide; ancorae small, angular, with acute apices; lophi lobiform, relatively narrow; lateral plates moderately large, with well-developed anterior projections.

Spermatheca: With relatively narrow, downcurved main reservoir and long diverticulum.

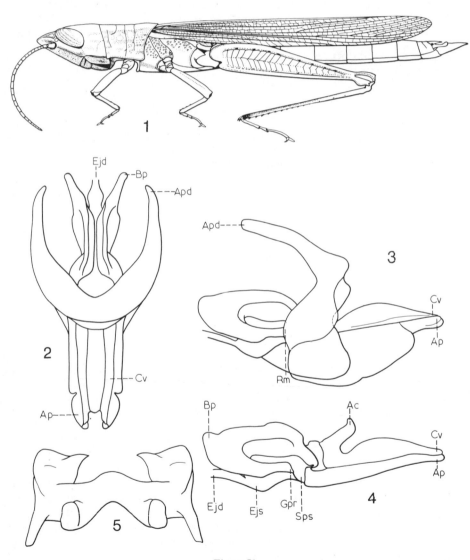

Figure 51.

1, *Tropidopola nigerica* Uvarov, 1937. Male. 2-5, phallic complex of *Tropidopola turanica* Uvarov, 1926. 2, phallic complex, dorsal view (membrane and epiphallus removed). 3, the same, lateral view. 4, endophallus, lateral view. 5, epiphallus.

Karyotype: Unknown.
Type genus: *Tropidopola* Stal, 1873.

The subfamily contains at present thirteen Old World genera. Possibly some of Neotropical genera ought to be referred to this subfamily.

Distribution: Eastern Mediterranean, Ethiopian Region, Oriental Region.

History of the subfamily is rather chequered. Stal (1877) placed genus *Tropidopola* in his 13th Division together with genera *Metapa* Stal, 1878; *Xenippa* Stal, 1878; *Gonyacanta* Stal, 1873; *Cervidia* Stal, 1878 and *Mesops* Serville, 1831. Brunner, 1882 placed *Tropidopola* into family *Opomalidae*. Jacobson and Bianki (1904) considered *Tropidopolae* of the subfamily *Catantopinae* and considered other present genera of the subfamily as a separate group — *Oxyrrhepes*. Dirsh (1961) restarted the group as a subfamily. Mistshenko (1963) regarded the group as a tribe of *Catantopinae*.

Finally the group was established as a subfamily by Dirsh (1965) containing at that time thirteen genera.

The genera of the subfamily are rather divergent which is indicating on their early separation. The nearest, but still rather remote affinity of *Tropidopolinae* may be suggested only with the subfamily *Leptacrinae*. Both of them possess closed mesosternal interspace, the same shape of body (which is suggestive but not decisive character) and certain similarity in the structure of the phallic complex.

Subfamily

Galideinae

(Fig. 52)

Diagnosis: Body of medium size, elongated stick-like, subcylindrical. Integument smooth or finely rugose. Head elongated, acutely conical; face, in profile, slightly incurved; fastigium of vertex elongated, acute angular; fastigial foveolae absent. Antennae filiform or slightly ensiform. Dorsum of pronotum subcylindrical; median and lateral carinae weak. Prosternal process present. Mesosternal interspace closed. Tympanum absent. Tegmina and wings shortened, vestigial or absent; venation of tegmen and wing not specialized. Hind femora slender; lower basal lobe shorter than upper one. Lobes of hind knee angular. External apical spine of hind tibia present. Male cercus relatively long, finger-shaped, at apex obtuse; supra-anal plate angular, at apex slightly attenuate; subgenital plate short, acutely conical. Ovipositor relatively long, valves straight at apices subacute.

Sound-producing mechanism not found.

PHALLIC COMPLEX: Ectophallus membraneous, except well differentiated cingulum; zygoma moderately small, apodemes long and slender, rami relatively large; valves of cingulum present. Endophallus strongly sclerotized; penis' sclerites divided on basal and apical valves, connected by short, very thin flexure; basal valves relatively large and slender; gonopore processes present; apical valves short and relatively robust. Epiphallus bridge-shaped; bridge long and narrow, widely divided in middle, with membraneous connection; small ancorae present or absent; lophi strong, incurved, hook-shaped; lateral plates small.

Spermatheca: With main reservoir rather long, narrow and curved; short diverticulum present.
Karyotype: Unknown.
Type genus: *Galideus* Finot, 1908.

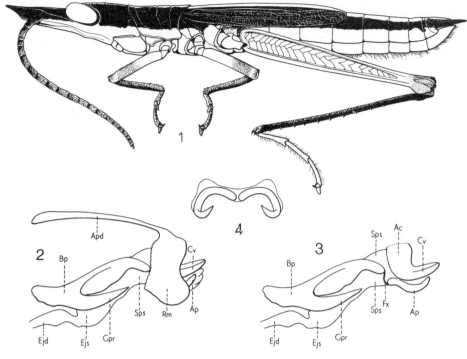

Figure 52.

1, *Galideus mocquerysi* (Finot, 1908). Male. 2-4, phallic complex of *Galideus elegans* Dirsh, 1962. 2, phallic complex, lateral view (membrane and epiphallus removed). 3, endophallus; lateral view. 4, epiphallus.

Besides the type genus, the following genera are referred to this subfamily: *Xenippacris* Descamps and Wintrebert, 1966; *Acutacris* Dirsh, 1966; *Xenippoides* Chopard, 1952.

Distribution: Madagascar.

The type genus *Galideus* was considered as a member of the subfamily *Hemiacridinae*. At present, on the basis of the phallic complex, which possess the flexure and divided epiphallus with hook-like lophi, it became necessary to erect a new subfamily for this group.

The new subfamily is probably related to the subfamilies *Hemiacridinae* and *Tropidopolinae*..

Subfamily

Opshomalinae

(Fig. 53)

Diagnosis: From medium to large size. Body elongated, cylindrical, stick-like. Head elongate, acutely conical; face, in profile, straight or slightly excurved; frontal ridge above and below antennae shallowly sulcate and carina-like between antennal scapes; fastigium of vertex elongate angular or short angular; fastigial foveolae absent. Antennae narrow ensiform. Dorsum of pronotum cylindrical; lateral carinae absent.

Prosternal process present. Mesosternal interspace open. Tympanum present. Tegmina and wings fully developed (rarely reduced) moderately narrow, venation longitudinal, straight intercalary vein of medial area absent, reticulation dense. Hind wings not specialized. Hind femora moderately narrow, lower basal lobe shorter than upper; upper lobe of hind knee rounded, lower lobe obtuse angular or rounded; hind tibia slightly widening towards distal end; external apical spine absent. Male cercus straight, narrow conical or with incurved apex; supra-anal plate angular; subgenital plate acute conical. Ovipositor short with valves curved at apices.

Sound-producing mechanism not detected.

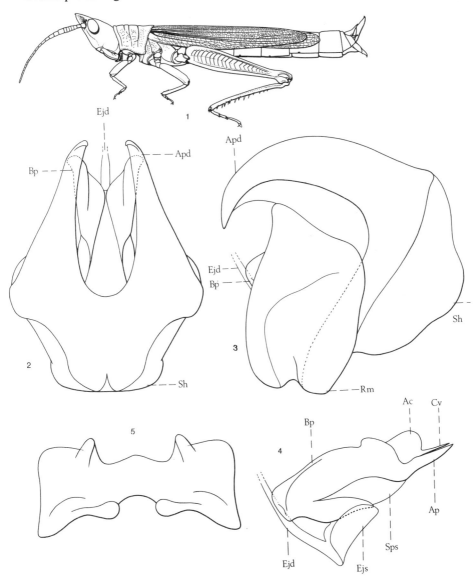

Figure 53.

1, *Opshomala viridis* Serville, 1831. 1, female. 2-5, phallic complex. 2, phallic complex, dorsal view (membrane and epiphallus removed). 3, the same, lateral view. 4, endophallus, lateral view. 5, epiphallus.

PHALLIC COMPLEX: Ectophallus in most parts sclerotized; cingulum very large well differentiated; zygoma large; apodemes very wide; rami narrow; large lateral sclerotizations present; valves of cingulum present. Endophallus strongly sclerotized; basal valves of penis long, robust, but relatively narrow; gonopore processes present; apical valves short, sturdy and almost straight; flexure almost of the same width as apical valves. Epiphallus bridge-shaped; bridge relatively wide; ancorae small, finger-shaped; lophi narrow lobiform; lateral plates large with apical projections almost not existing.

Spermatheca: Unknown.
Karyotype: Unknown.
Type genus: *Opshomala* Serville, 1831.

Only two genera of this subfamily can be recognized at present — *Opshomala* Serville, 1831 and *Copiocera* Burmeister, 1838.

The interrelation of the subfamily with the other subfamilies is obscure.

Subfamily

Oxyinae

(Fig. 54)

Diagnosis: Body of medium size or small, subcylindrical. Integument mostly smooth. Head subconical; face, in profile, straight or slightly excurved or incurved; frontal ridge shallowly sulcate; fastigium of vertex short, at apex rounded, parabolic or obtuse angular; fastigial foveolae absent. Antennae filiform. Dorsum of pronotum subcylindrical or flattened; median carina present or absent; lateral carinae absent. Prosternal process present. Mesosternal interspace open. Tympanum present. Tegmina and wings fully developed, shortened, lobiform or vestigial; venation of tegmen not specialized, reticulation dense; wing venation not specialized. Hind femora slender; lower basal lobe shorter than upper; upper lobe of hind knee rounded, lower lobe acute angular with sharp, spine-like apex. Hind tibia mostly widened towards apex; external apical spine present (except in genus *Gerista*). Distal sternites mostly with pair of rows brush-like hairs. Male cercus of various form; supra-anal plate angular or trilobate; subgenital plate short, subconical. Female subgenital plate with teeth or serration and of various form at apex. Ovipositor short, with slightly curved and on outer sides strongly serrated valves.

Sound-producing mechanism not found but in some genera (*Oxya*) its presence is suspected.

PHALLIC COMPLEX: Ectophallus membraneous with apical half and cingulum sclerotized, sometimes with strong lateral sclerotizations; cingulum well differentiated, with zygoma, apodemes and rami; valves of cingulum present, fused into plate. Endophallus well sclerotized; valves of penis connected by relatively long flexure; basal valves of penis relatively slender, diverging at proximal ends; gonopore processes present; apical valves relatively stout. Epiphallus bridge-shaped, sometimes asymmetrical; bridge short, divided or partly divided in middle; ancorae short, obtuse, or absent; lophi tooth-like, or hook-like; lateral plates moderately large.

Spermatheca: Main reservoir moderately wide, downcurved, with one relatively large diverticulum.
Karyotype: 2n ♂ = 23.
Type species: *Oxya* Serville, 1831.

This subfamily contains more than twenty known up to date genera.

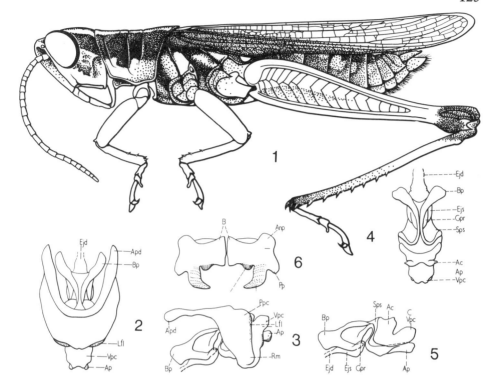

Figure 54.

1, *Oxya hyla* Serville, 1831. Male. 2-5, phallic complex (after Hollis, 1971). 2, phallic complex, dorsal view (membrane and epiphallus removed). 3, the same, lateral view. 4, endophallus, dorsal view. 5, the same, lateral view. 6, epiphallus.

Distribution: Africa, Asia, Austro-Asian Archipelago, Australia.

Oxyinae were established as a group *Oxyae* by Dirsh, 1956 and in 1961 raised by him to subfamily rank on the basis of characteristic phallic complex and on the basis of several external characters, such as spined lower lobe of hind knee, presence of hairs on the distal sternites, serrated ovipositor and toothed or serrated female subgenital plate. These characters are not exclusive features of the subfamily but are found in other subfamilies also, however, the comgination of them make the subfamily quite distinct.

Wide distribution of *Oxyinae* and the ecological niche they occupied — the humid habitat and in many genera semi-aquatic habit, indicate to their ancient origin. They are probably related to *Hemiacridinae* (*sensu lato*).

Subfamily

Coptacrinae

(Fig. 55)

Diagnosis: Body small or medium size, subcylindrical or laterally compressed. Integument smooth, slightly rugose or hairy. Head subconical, occiput mostly forming angle with vertex, which are separated by transverse ridge; fastigium of vertex short, of various shape; fastigial foveolae absent. Antenna filiform or slightly clavate. Dorsum

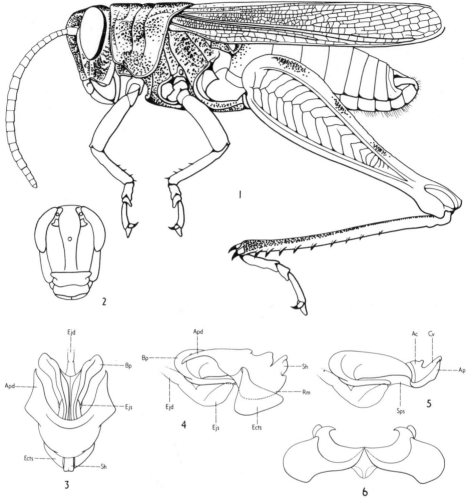

Figure 55.

1, *Eucoptacra poecila*, Uvarov, 1939. Male. 2, the same, face. 3-6, phallic complex. 3, phallic complex, dorsal view (membrane and epiphallus removed). 4, the same, lateral view. 5, endophallus, lateral view. 6, epiphallus.

of pronotum subcylindrical, tectiform or crested; lateral carinae absent. Prosternal process present. Mesosternal interspace open. Tympanum present. Tegmina and wings fully developed, shortened or vestigial; venation of tegmen and wing not specialized. Hind femora moderately slender; lower basal lobe shorter than upper one; lobes of hind knee rounded or angular. External apical spine of hind tibia absent. Last abdominal tergite in male mostly forming furcula. Male cercus of various shape; supra-anal plate mostly at apex attenuate; subgenital plate with transverse fold. Ovipositor short, valves moderately slender curved at apices or straight.

Sound-producing mechanism not found.

PHALLIC COMPLEX: Ectophallus membraneous with large lateral sclerotizations in distal part and fully differentiated cingulum, with large zygoma, wide apodemes and large, wide rami; valves of cingulum present. Endophallus strongly sclerotized; basal and apical valves of penis connected by flexure; basal valves of penis long and wide, at proximal ends expanded sideways; gonopore processes present, relatively long; apical valves of penis comparatively short and slender; both apical valves and valves of cingulum completely enclosed in sheath.' Epiphallus bridge-shaped; bridge fully or partly divided in middle; ancorae mostly present of various form; lophi lobiform of various size and form; lateral plates well developed.

Spermatheca: With narrow, twisted main reservoir and with one or two diverticula.
Karyotype: Unknown.
Type genus: *Coptacra* Stal, 1873.

Fourteen genera of this subfamily are known at present. Probably some of genera of the subfamily *Catantopinae* which were not sufficiently studied ought to be transferred to this subfamily.

Distribution: Tropical parts of Africa, Asia and Australia-Asian Islands.

The subfamily *Coptacrinae* was regarded by Brunner, 1893 as a group. Mistshenko, 1952 considered it as a tribe (but not in the same scope as Brunner). Dirsh, 1956 regarded it as a group, but in 1961 raised it to the subfamily rank.
The structure of the phallic complex, furcula in the last abdominal tergite, shape of supra-anal plate and presence of transverse fold in the male subgenital plate make it necessary to consider *Coptacrinae* as a separate subfamily.
This subfamily is probably related to the subfamily *Oxyinae* and to the subfamily *Hemiacridinae*, but true relationship is still not clear.

Subfamily

Calliptaminae

(Fig. 56)

Diagnosis: Body from small to large size, sturdy, subcylindrical. Integument smooth or finely rugose. Head subconical or subglobular; face, in profile, straight or slightly excurved; fastigium of vertex short; fastigial foveolae absent. Antennae filiform. Dorsum of pronotum flattened or weakly tectiform; median and lateral carinae present. Prosternal process present. Mesosternal interspace open. Tympanum present. Tegmina and wings fully developed or shortened; venation of tegmen and wing not specialized. Hind femora mostly widened, with lower basal lobe shorter than upper one; lobes of hind knee rounded; external apical spine of hind tibia absent. Male cercus large, strong, forceps-like; supra-anal plate angular; subgenital plate subconical. Ovipositor short, with valves curved at apices.
Sound-producing mechanism not found.

PHALLIC COMPLEX: Ectophallus partly membraneous forming very large cingulum and lateral sclerotizations; zygoma very large, apodemes short; rami moderately large; valves of cingulum present. Endophallus strongly sclerotized; penis' sclerites divided on basal and apical valves connected by flexure; basal valves of penis large and sturdy at proximal ends widely spread sideways; gonopore processes present; apical valves relatively short and wide. Epiphallus disc-shaped; ancorae short, mostly finger-shaped, sometimes placed in middle part of disc; lophi absent; lateral plates narrow.

Spermatheca: With narrow, downcurved main reservoir and one diverticulum.
Karyotype: Unknown.
Type genus: *Calliptamus* Serville, 1831.

The subfamily at present contains about twenty genera.

128

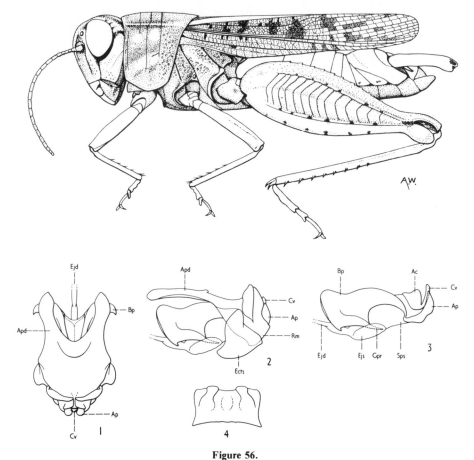

Figure 56.

Calliptamus italicus (Linnaeus, 1758). Male. 2-4, phallic complex of *Acorypha nigrovariegata* (I. Bolivar, 1889). 1, phallic complex, dorsal view (membrane and epiphallus removed). 2, the same, lateral view. 3, endophallus, lateral view. 4, epiphallus.

Distribution: Europe, Africa, Asia, Madagascar.

Subfamily *Calliptaminae* was erected by Jacobson and Bianki in 1904, but later authors regarded them as a part of the subfamily *Catantopinae*. Mistshenko, 1952, treated them as a tribe of *Catantopinae*. They were restored as the subfamily by Dirsh, 1956.

By their mobile cerci, adapted for grasping and characteristic discoidal epiphallus *Calliptaminae* differs from the other subfamilies of *Acrididae*. A remote affinity with subfamily *Eyprepocneminae* may be suggested.

Subfamily

Euryphyminae

(Fig. 57)

Diagnosis: Body from medium to small size, subcylindrical. Integument mostly strongly or slightly rugose. Head subconical; face, in profile, straight or slightly excurved; fastigium of vertex short, mostly angular, fastigial foveolae absent. Antennae filiform. Dorsum of pronotum of various shape; median and lateral carinae present. Prosternal process present. Mesosternal interspace open. Tympanum present. Tegmina and wings fully developed, shortened or vestigial; venation of tegmen and wing not specialized. Hind femora mostly wide; lower basal lobe shorter than upper one. Lobes of hind knee rounded. External apical spine of hind tibia absent. Posterior margin of last abdominal tergite in male strongly sclerotized, toothed or serrated. Male cerci of various and sometimes very complicated shape, highly mobile, with large basal articulation; supra-anal plate transverse or elongate, mostly with complicated sculpture; subgenital plate short, subconical. Ovipositor short moderately sturdy; valves slightly curved at apices.

Sound-producing mechanism not found.

PHALLIC COMPLEX: Ectophallus partly membraneous, with various form of sclerotizations and very large cingulum; zygoma large, elongate, apodemes relatively short; rami moderately large; valves of cingulum present. Endophallus strongly sclerotized; penis' sclerites divided on basal and apical valves connected by stout

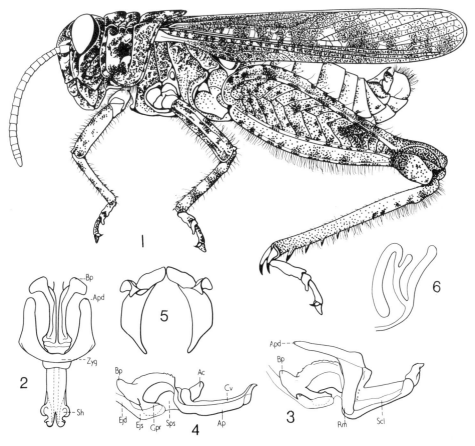

Figure 57.

1, *Euryphymus haematopus* (Linnaeus, 1758). Male. 2-6, phallic complex and spermatheca of *Phymeurus rhodesianus* Mason, 1966. 2, phallic complex, dorsal view (membrane, ectophallic sclerotization and epiphallus removed). 3, the same, lateral view. 4, endophallus, lateral view. 5, epiphallus. 6, spermatheca. (After Mason, 1966).

flexure; basal valves of penis large, widened, at proximal ends diversed sideways; gonopore processes present; apical valves of penis relatively long and slender, together with valves of cingulum covered with sheath. Epiphallus bridge-shaped, with bridge short and deeply divided in middle; short ancorae present; lophi very large, acute angularly lobiform or roundly lobiform; lateral plates small or strongly reduced.

Spermatheca: Moderately large, of oval shape main reservoir downcurved. One or two curved diverticula present.
Karyotype: Unknown.
Type genus: *Euryphymus* Stal, 1873.

More than twenty genera of this subfamily are known at present.

Distribution: South, West and East Africa.

Subfamily *Euryphyminae* were regarded as members of the subfamily *Catantopinae*. In 1956 Dirsh erected for them new subfamily, on the basis of peculiar structure of their phallic complex, of their strongly developed articulation of male cercus and strong sclerotization of the last abdominal tergite of male.

The interrelations of *Euryphyminae* with the other subfamilies of *Catantopidae* is not clear. The only thing can be said that they probably belong to *Catantopidae* not having close alliance with other subfamilies.

Subfamily

Eyprepocneminae

(Fig. 58)

Diagnosis: Body from large to small, subcylindrical. Integument smooth or finely rugose. Head subconical or subglobular; face, in profile, straight or slightly excurved; fastigium of vertex short, rounded or slightly angular; fastigial foveolae absent. Antennae of various shape. Dorsum of pronotum flattened or slightly tectiform; median carina present; lateral carinae mostly present. Prosternal process present. Mesosternal interspace open. Tympanum present. Tegmina and wings fully developed or shortened; venation of tegmen and wing not specialized. Hind femora mostly slender; lower basal lobe shorter than upper one; lobes of hind knee rounded or obtusely angular. External apical spine of hind tibia absent. Male cerci of various shape; supra-anal angular; subgenital plate short, subconical. Ovipositor short, sturdy; valves curved at apices.

Sound-producing mechanism not found.

PHALLIC COMPLEX: Ectophallus membraneous mostly with large lateral sclerotizations; cingulum large, strongly sclerotized; zygoma of various size, apodemes relatively short, rami of various size; valves of cingulum present. Endophallus strongly sclerotized; penis' sclerites divided on basal and apical valves, connected by long flexure; basal valves of penis wide far spreading sideways at proximal ends; gonopore processes present; apical valves of penis relatively long, slender, often together with valves of cingulum covered with sheath. Epiphallus bridge-shaped; bridge mostly wide and partly membraneous; ancorae large, curved, articulated with lateral plates; lophi large, lobiform; lateral plates small.

Spermatheca: With narrow, downcurved main reservoir and one diverticulum.
Karyotype: Unknown.
Type genus: *Eyprepocnemis* Fieber, 1853.

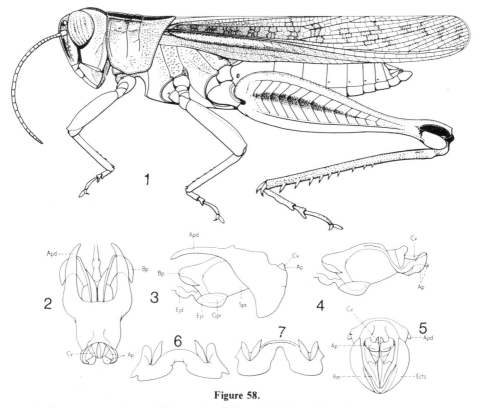

Figure 58.

1, *Eyprepocnemis plorans* (Charpentier, 1825), male. 2-7, phallic complex. 2, dorsal view, (epiphallus and ectophallic membrane removed). 3, the same, lateral view. 4, endophallus, lateral view. 5, distal end of phallic complex, posterior view. 6, epiphallus, bridge in horizontal position. 7, the same, bridge in vertical position.

This subfamily contains more than 30 known genera and probably some of the genera now referred to *Catantopinae* ought to be transferred to this subfamily.

Distribution: South Palearctic, Ethiopian, Oriental, Malagassi and Australian Regions.

Subfamily *Eyprepocneminae* was erected by Jacobson & Bianki, 1904. Later authors considered them as group *Eyprepocnemi* of the subfamily *Catantopinae*. Mistshenko, 1952 regarded them as a tribe. Dirsh, 1956 considered them as a group and later, in 1961, raised it to subfamily rank.

This subfamily is rather well defined by its phallic complex, which despite wide variability still constitute a good character.

Eyprepocneminae are related to certain extent to subfamily *Calliptaminae*.

Subfamily

Cyrtacanthacrinae

(Fig. 59)

Diagnosis: Body from very large to medium size, subcylindrical. Integument smooth or finely rugose. Head subconical or subglobular; face, in profile, straight or slightly

132

excurved; fastigium of vertex short; fastigial foveolae absent. Antennae filiform. Dorsum of pronotum tectiform or slightly saddle-shaped; lateral carinae absent. Prosternal process present. Mesosternal interspace open; mesosternal lobes rectangular. Tympanum present. Tegmina and wings fully developed or shortened; venation of tegmen and wing not specialized. Hind femora moderately slender; lower basal lobe shorter than upper one; lobes of hind knee rounded or obtusely angular. External apical spine of hind tibia absent. Male cercus of various shape; supra-anal plate angular; subgenital plate conical or subconical. Ovipositor short and sturdy, with valves curved at apices.

Sound-producing mechanism not found.

PHALLIC COMPLEX: Ectophallus membraneous, except lateral sclerotizations and heavily sclerotized well differentiated cingulum; zygoma relatively narrow, apodemes short, rami relatively large; valves of cingulum present, sometimes strongly reduced and sometimes fused. Endophallus heavily sclerotized; sclerites of penis divided on basal and apical valves, connected by thick flexure; basal valves of penis wide, at proximal ends widely expanded sideways; gonopore processes present; apical valves of penis relatively slender, at apices of various, sometimes complicated, shape; and sometimes, together with valves of cingulum enclosed in sheath. Epiphallus bridge-shaped, bridge narrow; ancorae, if present, small, mostly absent; lophi large, lobiform; lateral plates small.

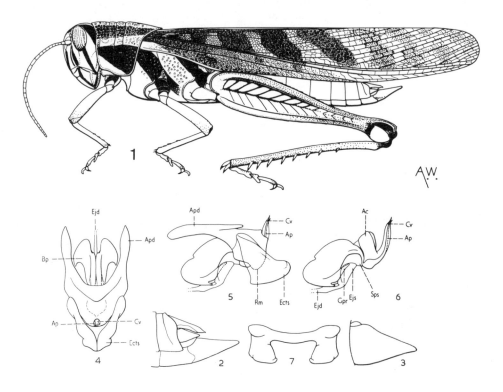

Figure 59.

1, *Ornithacris cyanea*, male. 2, end of male abdomen, lateral view. 3, male cercus. 4-6, phallic complex: 4, dorsal view, (epiphallus and most part of ectophallic membrane removed). 5, the same, lateral view. 6, endophallus, lateral view. 7, epiphallus, lophi in semi-vertical position.

Spermatheca: Main reservoir of oval form, downcurved, with one comparatively long, twisted diverticulum.
Karyotype: 2n ♂ = 23. Based on few observations.
Type genus: *Cyrtacanthacris* Walker, 1870.

About 30 genera of this subfamily are known presently.

Distribution: Tropical, subtropical and partly temperate zones of Whole World.

The name *Cyrtacanthacrinae* or *Cyrtacanthacridinae* was used previously as group, tribe or subfamily name instead of name *Catantopinae*, particularly by American authors, however, the genus *Catantops* was described by Schaum in 1853 and genus *Cyrtacanthacris* by Walker in 1870. So usage of the names was incorrect. At present, when *Catantopinae* and *Cyrtacanthacrinae* are regarded as the separate subfamilies, the correct usage of the subfamily names became obvious.

The only good external character separating *Cyrtacanthacrinae* from *Catantopinae* and other subfamilies, is the rectangular form of mesosternal lobes, which in other subfamilies are rounded. The other characters are thick flexure and often complicated apices of apical valves of penis.

Probably this subfamily is closely related to the *Catantopinae* and particularly to the group *Catantopini.*

Subfamily

Podisminae

(Fig. 60)

Diagnosis: Body stout subcylindrical, of medium or small size. Head short, from obtusely conical to subglobular; face, in profile, straight, slightly inclined backwards; frontal ridge low, flat or shallowly concave, mostly without lateral carinulae; fastigium of vertex short, at apex rounded or widely obtuse angular; fastigial foveolae absent. Antennae filiform. Dorsum of pronotum subcylindrical, slightly flattened, or slightly saddle-shaped, lateral carinae absent. Prosternal process present. Mesosternal interspace open, mostly wide. Tympanum present, well developed, strongly reduced or absent. Tegmina and wings from fully developed to strongly reduced, sometimes absent; membrane and reticulation coarse, venation not specialized. Hind femora from comparatively slender to wide and stout; with lower basal lobe shorter than upper; lobes of hind knee rounded. Hind tibia straight, or slightly curved; external apical spine absent. Male cerci and supra-anal plate of various shape; subgenital plate subconical. Ovipositor short, with curved valves.
Sound-producing mechanism not found.

PHALLIC COMPLEX: Ectophallus membraneous with strongly sclerotized cingulum; zygoma relatively short; apodemes wide and sturdy; rami elongated; apical part of ectophallus of various shape, strongly sclerotized; valves of cingulum present. Endophallus strongly sclerotized; basal valves of penis sclerites large, widened and excurved sideways at proximal ends; gonopore processes present; flexure short and narrow; apical valves relatively wide. Epiphallus bridge-shaped; bridge moderately narrow; ancorae short, angular; lophi lobiform, short; lateral plates well developed; apical projections small.

Spermatheca: With oval, downcurved main reservoir and vermiform one or two diverticula.

134

Karyotype: $2n\,\male = 12, 21, 23$. In majority of studied cases $2n\,\male = 23$.
Type genus: *Podisma* Berthold, 1827.

Numerous genera and large amount of species belong to this subfamily.

Distribution: Palearctic, Nearctic and Neotropical Regions.

The subfamily probably originated from the *Catantopinae* stock comparatively recently and probably in Old World. In warm climatic condition in the early Tertiary they penetrated across the Beringia Land into N. America where they radiated into large number of genera and species. The next step was when they distributed along the Andes up to the southern part of South America, and even to the Galapagos Islands.

The subfamily as a whole has rather disrupted pattern of distribution. Some genera and species have insular characteristic in the distribution. Being far isolated from the other genera and remains as relic after glaciation. They are usually connected with mountainous landscapes.

The oldest genus of the subfamily genus *Podisma*, which was described by Berthold in 1827. Accordingly the Old World genera and species were first recognised as a group

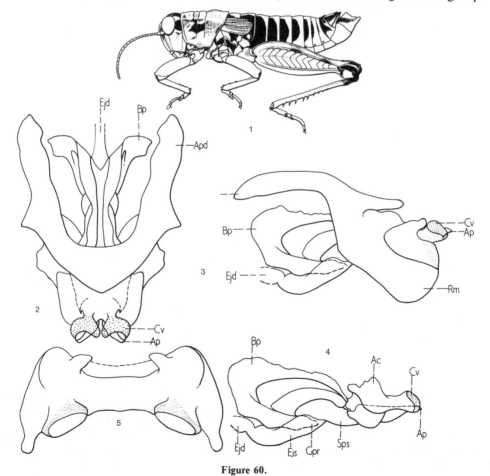

Figure 60.

1, *Podisma pedestris* (Linnaeus, 1758). Male. 2-5, phallic complex. 2, phallic complex, dorsal view, (membrane and epiphallus removed). 3, the same, lateral view. 4, endophallus, lateral view. 5, epiphallus.

of subfamily rank by Jacobson and Bianki in 1904. Later the group was treated as a tribe. Brunner von Wattenwyl, 1893 placed the tribe into mixed group *Pezotettiges*. Willemse, 1921 regarded it as part of his group *Tonkinacridae*. Tinkham, 1940 regarded them as group *Podismae*. Mistshenko in 1947 and 1952 restored them to rank of tribe, and here it is considered as a subfamily. The features of the subfamily is not very definite, but the combination of the characters justified to consider them at that level.

American representatives of the subfamily are traditionally called by the American authors as a group *Melanopli*. However, genus *Melanoplus* was described by Stal in 1873 and was rightly included by Mistshenko, 1952 into tribe *Podismini*, which has priority for the name.

By the shape of body and by the many common characters in the structure of genitalia, the *Podisminae* are closely related to the subfamily *Catantopinae*. They probably branched from this subfamily comparatively recently, probably in time when forests began to decrease and shrubs and grasses began to dominate the earth landscapes.

Subfamily

Catantopinae

(Fig. 61)

Diagnosis: Body from large to small size, and various shape. Head of various shape; fastigial foveolae absent. Antennae mostly filiform. Dorsum of pronotum of various shape; median carina present or absent; lateral carinae mostly absent. Prosternal process present. Mesosternal interspace open. Tympanum mostly present. Tegmina and wings fully developed, shortened, lobiform, vestigial or absent; venation of tegmen and wing not specialized. Hind femora slender or widened; lower basal lobe usually shorter than upper one. Lobes of hind knee rounded or angular. External apical spine of hind tibia present or absent. Male cercus, supra-anal plate and subgenital plate of various shape. Ovipositor of various shape.

Sound-producing mechanism not found.

PHALLIC COMPLEX: Ectophallus membraneous or partly sclerotized; cingulum well developed and well differentiated; valve of cingulum present. Endophallus well sclerotized; penis' sclerites divided to basal and apical valves, connected by flexure; basal valves of penis usually wide; gonopore processes present; apical valves of various shape, sometimes covered with sheath together with valves of cingulum. Epiphallus bridge-shaped of various form, sometimes bridge divided or semi-divided in middle; ancorae mostly present, of various form; lophi present of various shape.
Spermatheca: Of various shape.
Karyotype: $2n \, \male = 12, 19, 21, 22, 23$. Predominantly $2n \male = 23$.
Type species: *Catantops* Schaum, 1853.

This highly heterogeneous subfamily at present contains more than one hundred genera. It needs to be split on several subfamilies or divided into tribes. Regretfully, in the present state of our knowledge of the group, it is not possible to offer good scheme for the division. According to the estimation of the present author, probably only one-third of the genera of this subfamily is known.

Distribution: Tropical and subtropical zones of the Whole World.

Relationship of the subfamily with the other subfamilies of *Acrididae* is extremely complicated. Owing to the diversity in external appearance, diversity of the phallic

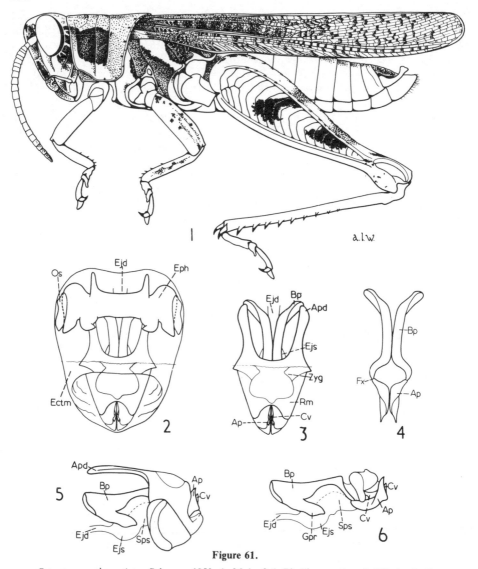

Figure 61.

Catantops melanostictus Schaum, 1853. 1. Male. 2-6. Phallic complex. 2. Whole phallic complex from above. 3. The same, but ectophallic membrane and epiphallus removed. 4. Penis, from above. 5. As fig. 3, but in profile. 6. Endophallus, in profile.

complex and wide range of the ecological niches occupied by the genera of this subfamily, it is possible to link them with almost every subfamily of *Acrididae*. However, only when this large assemblage of genera will be better studied and more of them will be discovered and described, only then will it be possible to classify them properly, with more reliable scientific background. In the first half of this century most of the subfamilies of present *Acrididae* were regarded as members of *Catantopinae*. Now these subfamilies—*Coptacrinae, Calliptaminae, Dericorythinae, Euryphyminae, Eyprepocneminae, Hemiacrinae, Oxyinae, Lithidiinae, Romaleinae* etc. including the new subfamilies erected in this work, are eliminated from *Catantopinae*. In spite of this reduction of the subfamily, it still needs to be reclassified when more material will be available.

Subfamily

Pargainae

(Fig. 62)

Diagnosis: Body elongated, small or of medium size. Head acutely conical; face straight or slightly incurved; fastigium of vertex large, elongate, at apex mostly rounded; fastigial foveolae absent. Antennae ensiform. Dorsum of pronotum flat;

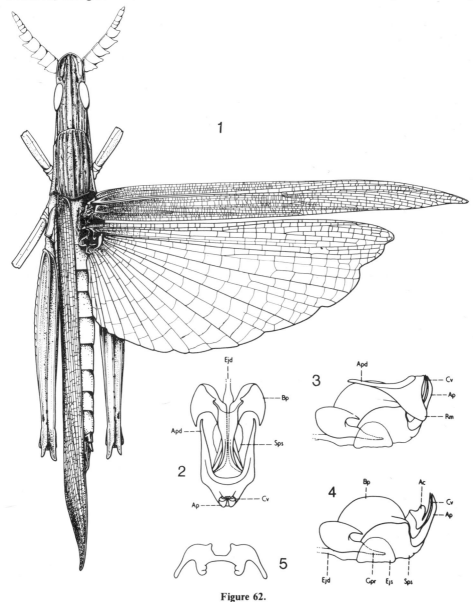

Figure 62.

1, *Parga xanthoptera* (Stal, 1855). Male. 2-5, phallic complex of *Parga taeniata* (I. Bolivar, 1889). 2, phallic complex, dorsal view (membrane and epiphallus removed). 3, the same, lateral view. 4, endophallus, lateral view. 5, epiphallus.

median and lateral carinae present. Prosternal tubercle present. Mesosternal interspace open. Tegmina and wings fully developed, or shortened; venation dense, with numerous longitudinal intercalary veins; intercalary vein of medial area mostly present. Hind femora mostly relatively narrow. Hind knee mostly with elongated acute lobe or lobes.

Stridulatory mechanism not found.

PHALLIC COMPLEX: Ectophallus membraneous, except sclerotized cingulum; valves of cingulum present. Endophallus strongly sclerotized. Basal valves of penis relatively very large and wide; widely spreading at proximal ends; gonopore processes present; apical valves relatively slender, upcurved; flexure long and narrow. Epiphallus bridge-shaped, bridge narrow, ancorae absent or short, rudimentary; lophi short, mostly bilobate; lateral plates well developed.

Spermatheca: Main reservoir oval, relatively large, with small apical diverticulum.
Karyotype: Unknown.
Type genus: *Parga* Walker, 1870.

Distribution: Ethiopian Region, Madagascar.

The other studied genera of the subfamily:

> *Acteana* Karsch, 1896.
> *Amphicremna* Karsch, 1896.
> *Machaeridia* Stal, 1873.
> *Paraparga* I. Bolivar, 1909.
> *Pargaella* I. Bolivar, 1909.
> *Pseudopargaella* Descamps & Wintrebert, 1966.

The main difference of *Pargainae* from other subfamilies of *Acridoidea* is a very characteristic structure of epiphallus. The other characters — dense venation with numerous intercalary veins, and presence of prosternal tubercle, *Pargainae* share with many other unrelated genera of different subfamilies. The combination of these characters however, define the subfamily clearly.

The subfamily is transferred into family *Catantopidae* on the basis of the sum of characters.

Subfamily

Paraconophyminae

(Fig. 63)

Diagnosis: Body small, short and sturdy, subcylindrical. Head short, obtusely conical; face, in profile, straight, slightly oblique; frontal ridge flat or slightly convex, without lateral carinulae; fastigium of vertex short, obtuse angular; fastigial foveolae detectable. Antennae short, filiform. Dorsum of pronotum subcylindrical; median and lateral carinae present. Prosternal process or tubercle present. Mesosternal interspace open, wide. Tympanum well developed. Tegmina and wings strongly reduced, lateral lobiform. Hind femora short and sturdy, with lower basal lobe shorter than upper one; lobes of hind knee rounded. Hind tibia slightly widened towards distal end; external apical spine present. Male supra-anal plate and cerci of various form. Subgenital plate short, subconical. Ovipositor short with valves curved at apices.

Sound-producing mechanism not found.

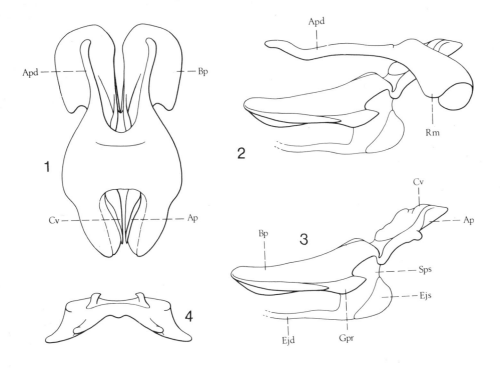

Figure 63.

Dinaria mirzoyani Popov, 1951. Phallic complex. 1, dorsal view (membrane and epiphallus removed). 2, the same, lateral view. 3, endophallus, lateral view. 4, epiphallus.

PHALLIC COMPLEX: Ectophallus membraneous, partly sclerotized at apical end; cingulum differentiated; zygoma large; apodemes relatively short and slender; rami narrow; valves of cingulum present. Endophallus strongly sclerotized; basal valves of penis sclerites relatively long, at proximal ends considerably spreading sideways; gonopore processes present; flexure short and sturdy; apical valves of penis short, upcurved. Epiphallus bridge-shaped, incurved, at apices acute; lophi short, lobiform, with obtuse apices; lateral plates large, with anterior projections well developed.

Spermatheca: With oval, downcurved main reservoir and one short diverticulum.
Karyotype: Unknown.
Type genus: *Paraconophyma* Uvarov, 1921.

Only two genera were examined: the type genus and *Dinaria* Popov, 1951.

Distribution: Kashmir, Afghanistan, Iran.

Interrelation of this subfamily with the other subfamilies of *Acridoidea* is obscure. The only definite conclusion can be drawn that they do not relate to the subfamily *Conophyminae*. The similar external appearance of both subfamilies was the reason to regard them as one group. However, the structure of the phallic complex is quite different indicating that they are not related even remotely.

Shelforditinae

(Fig. 64)

Diagnosis: Body small or medium size, subcylindrical. Integument rugose. Head hypognathous or slightly prognathous; fastigium of vertex short, sloping forwards; fastigial furrow absent; fastigial foveolae absent. Antennae filiform. Dorsum of pronotum subcylindrical. Prosternal tubercle present. Mesosternal interspace open or closed. Tegmina, wings and tympanum absent. Hind femora from slender to moderately wide; lower basal lobe shorter than upper one. External apical spine of hind tibia absent. Two last distal tergites of abdomen divided along middle, with large,

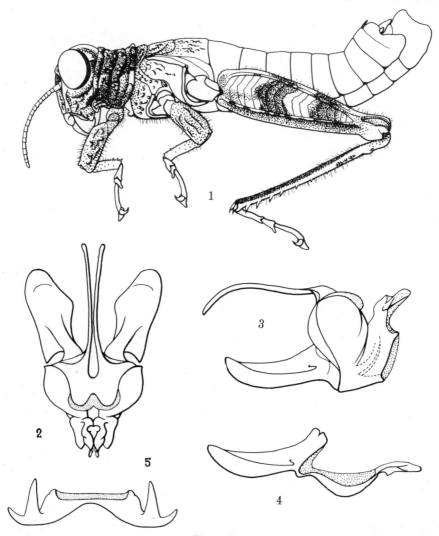

Figure 64.

1, *Shelfordites aberrans* Karny, 1910. Male. 2-5, phallic complex. 2, phallic complex, dorsal view. 3, the same, lateral view. 4, endophallus, lateral view. 5, epiphallus (figs. 2-5, after Brown, 1967).

bifurcate appendage protruding from this slit (in male, in female the tergites are divided, but appendage absent). Male cercus simple short subconical; supra-anal plate with deep transverse furrow in middle; subgenital plate rounded or conical. Ovipositor short, with valves slightly curved.

Sound-producing mechanism not found.

PHALLIC COMPLEX: Ectophallus membraneous except slightly sclerotized distal part and strongly sclerotized cingulum; zygoma, apodemes and rami present; valves of cingulum absent. Endophallus strongly sclerotized; sclerites of penis divided on basal and apical valves which are connected by short flexure; basal valves of penis wide, expanded sideways at proximal ends; gonopore processes absent; apical valves of penis relatively slender, enclosed into membraneous sheath. Epiphallus bridge-shaped; bridge narrow; ancorae short, finger-shaped; lophi large, hook- or finger-shaped, upcurved. Oval sclerites present.

Spermatheca: Not known.
Karyotype: Unknown.
Type genus: *Shelfordites* Karny, 1910.

Besides the type genus, the two genera of this subfamily are known — *Kalaharicus* Brown, 1960 and *Occidentula* Brown, 1967.

Shelforditinae up to present time were considered as the genera of the heterogeneous family *Lentulidae*. However, the flexured valves of penis and absence of sclerotized semi capsule-like of distal half of ectophallus not allowed to keep them in the family, here they are transferred as subfamily into the family *Catantopidae*.

The interrelation of this subfamily with other subfamilies cannot be suggested in the present knowledge of *Acridoidea*.

Subfamily

Illapelinae

(Fig. 65)

Diagnosis: Body exceptionally small, subcylindrical, with integument rugose. Head subconical unproportionally large; face, in profile, almost straight, slightly inclined; frontal ridge shallowly sulcate; fastigium of vertex short, wide, at apex widely rounded fastigial foveolae absent. Antennae very short five-six segmented, scape long, apical segment elongated, club-like. Dorsum of pronotum short, flattened without carinae. Prosternum low elevated. Mesosternal interspace wide, open. Tympanum absent. Tegmina and wings absent. Hind femur short and wide; lower basal lobe shorter than upper. Lobes of hind knee obtuse angular. External apical spine of hind tibia absent. Male cercus short, angular; supra-anal plate relatively wide angular; subgenital plate short, subconical. Ovipositor short, slender, almost straight.

Sound-producing mechanism not detected.

PHALLIC COMPLEX: Ectophallus membraneous except sclerotized cingulum; zygoma of cingulum wide bar-like; apodemes short and sturdy, parallel; rami large, articulated with large ventral sclerotizations; valves of cingulum absent. Endophallus relatively small; sclerites of penis divided, basal part of valves slender; gonopore processes absent; apical part covered with wide sheath flexure very short. Epiphallus bridge-shaped; bridge relatively narrow; ancorae relatively large, angular; lophi angularly lobiform; lateral plates relatively large, with large apical projections.

Spermatheca: Simple, tube-like, curved, at apex slightly widened.

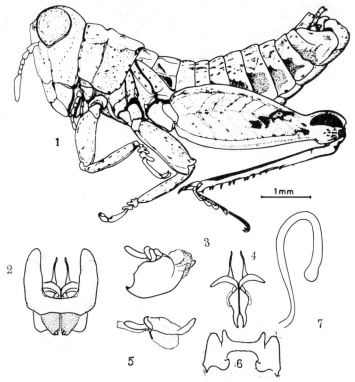

Figure 65.

1, *Illapelia penai* Carbonell and Mesa, 1972. Male. 2-6, phallic complex. 2, phallic complex, dorsal view (membrane and epiphallus removed). 3, the same, lateral view. 4, endophallus, dorsal view. 5, the same, lateral view. 6, epiphallus. 7, spermatheca. (After Carbonell and Mesa, 1972).

Karyotype: $2n\sigma = 23$.

Type genus: *Illapelia* Carbonell & Mesa, 1972.

Only one genus of this subfamily is known.

Distribution: Chilean Andes, 2,800-3,000 metres altitude.

This new remarkable subfamily was recently erected by Carbonell and Mesa. As the authors stated *Illapelinae* cannot be fitted into any known subfamily and its affinities with the other subfamilies of the *Acridoidea* cannot be even remotely suggested.

The habitat of the subfamily in mountains at 2,800-3,000 metre altitude suggests that they have very short time for their life cycle. As a result, they are probably neotenic. It can be supported from morpholigical point of view by the disproportionately large head, by absence of tegmina, wings and tympanum and by general nymphal appearance.

Illapelinae are the smallest known *Acridids* — male being 5.1-5.3 and female 8.1-9.3 mm.

Family

Acrididae

Diagnosis: Body of various size and shape. Head of various shape; fastigial furrow absent; fastigial foveolae present or absent. Antennae of various form. Dorsum of pronotum of various shape, mostly with median and lateral carinae. Prosternal process, tubercle or collar absent, (rarely present). Mesosternal interspace open or closed. Tympanum present or absent. Tegmina and wings fully developed, shortened, lobiform, vestigial or absent. Lower basal lobe of hind femur shorter or same length as upper one. External apical spine of hind tibia mostly absent.

Sound-producing mechanism, if present, mostly of tegmino-femoral type.

PHALLIC COMPLEX: Ectophallus membraneous or partly sclerotized; well developed cingulum present; valves of cingulum present. Endophallus well sclerotized; sclerites of penis, divided on basal and apical valves connected by flexure. Gonopore processes present. Spermatophore sac in middle position. Epiphallus bridge-shaped, with ancorae and lophi. Oval sclerites present.

Spermatheca: Of various shape.
Type genus: *Acrida* Linnaeus, 1758.

Distribution: Whole World.

List of subfamilies

1. *Acridinae*
2. *Chrysochraontinae*
3. *Eremogryllinae*
4. *Gomphocerinae*
5. *Gymnobothrinae*
6. *Hyalopteryxinae*
7. *Oedipodinae*
8. *Phlaeobinae*
9. *Truxalinae*

Key to subfamilies

1 (14) Sound-producing mechanism present.

2 (7) Sound-producing mechanism represented by row articulated stridulatory pegs on inner side of hind femur.

3 (4) Fastigial foveolae absent.
                                                    Chrysochraontinae

4 (3) Fastigial foveolae present.

5 (6) Epiphallus divided or semi-divided in middle.
                                                    Eremogryllynae

6 (5) Epiphallus not divided in middle.
                                                    Gomphocerinae

7 (2) Sound-producing mechanism of various structure, but without articulated pegs.

8 (11) Sound-producing mechanism of femora-tegminal type. Speculum on hind wing absent.

9 (10) Sound-producing represented by unarticulated serration on inner side of hind femur and sharpened radial and medial veins on tegmina.
                                                    Truxalinae

10 (9) Sound-producing mechanism represented by sharp ridge on inner side of hind femur and serrated intercalary vein in medial area of tegmen.

Oedipodinae

11 (8) Sound-producing mechanism of wing-tegminal type. Speculum on hind wing present. Veins on inner side of tegmen sharp.

12 (13) Wing speculum located in medial area of wing.

Acridinae

13 (12) Wing speculum located in first cubital area of wing.

Hyalopteryxinae

14 (1) Sound-producing mechanism not found.

15 (16) Fastigial foveolae mostly present. Epiphallus with large finger-shaped lophi, hooked at apices.

Gymnobothrinae

16 (15) Fastigial foveolae absent. Epiphallus with lobiform lophi.

Phlaeobinae

Subfamily

Chrysochraontinae

(Fig. 66)

Diagnosis: Body from strongly elongate, narrow cylindrical, to short subcylindrical; size from large to small. Head from strongly elongate, acutely conical, short acutely conical, or obtusely conical; face, in profile, straight or slightly incurved; fastigium of vertex from elongated to short, at apex angular or rounded; fastigial foveolae mostly absent. Antennae ensiform or narrow ensiform. Dorsum of pronotum flat, slightly tectiform, subcylindrical or slightly saddle-shaped. Prosternal process or tubercle mostly absent. Tegmina and wings fully developed, shortened or rarely absent; reticulation sparse; intercalary vein of medial area of tegmen absent. Hind wings without specialization. Hind femora slender or moderately widened. Knees of hind femur with rounded or obtuse angular lobes. Hind tibia not specialised; external apical spine absent. Male cercus mostly simple, conical; supra-anal plate angular; subgenital plate elongated acutely conical, conical or subconical. Ovipositor mostly short, with valves curved at apices or straight.

Stridulatory mechanism represented by serration on inner side of hind femora consisting of row of articulated pegs forming straight, regularly undulated, or irregular line. Pegs sometimes placed very densely, sometimes sparsely; at the ends of row pegs are always more sparse than in middle. Pegs sometimes connected with ridge by membraneous connection (articulated pegs) or their bases enclosed into follicles. Sound-producing is achieved by rubbing row of pegs against sharp, convex radial and medial veins of tegmen.

PHALLIC COMPLEX: Ectophallus membraneous except strongly sclerotized cingulum; zygoma narrow; apodemes long, forming widely arcuate structure; rami wide; valves of cingulum present. Endophallus strongly sclerotized; basal valves of penis large and wide, spreading sideways at proximal ends; gonopore processes present; apical valves of penis moderately narrow, upcurved; flexure relatively long and narrow. Epiphallus bridge-shaped; bridge short and narrow; ancorae short, acute at apices, articulated

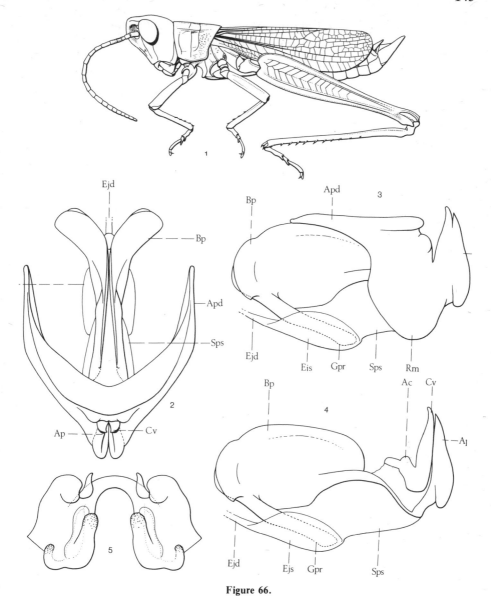

Figure 66.

1, *Chrysochraon dispar* (Germar, 1831-1835). Male. 2-5, phallic complex. 2, phallic complex, dorsal view, (membrane and epiphallus removed). 3, the same, lateral view. 4, endophallus, lateral view. 5, epiphallus.

with ends of bridge; lophi bilobate, monolobate or sometimes trilobate attached to pair of branches extending from the basal sides of bridge; lateral plates from narrow to wide, with angular posterior projections.

Spermatheca: With large, downcurved, approximately oval, reservoir and narrow, finger-shaped diverticulum.

Karyotype: 2n♂=17, 23. Predominantly 2n♂=23.

Type genus: *Chrysochraon* Fischer, 1853.

Distribution: All zoogeographical Regions, except Australia.

Subfamily *Chrysochraontinae* are near to *Gomphocerinae* but differs by ensiform or narrow ensiform antennae, absence of fastigial foveolae or if present, they are of different type, and by the karyotype. In majority of investigated *Chrysochaontinae* karyotype is $2n \male = 23$, while in *Gomphocerinae* in great majority of investigated cases, karyotype is $2n \male = 17$.

Chrysochraontinae were established as a subfamily by Jacobson & Bianki (1904) for rather mixed and unrelated group of genera (*Chrysochraon, Duronia, Parapleurus, Platypterna*). The only character which united them was absence of fastigial foveolae and ensiform antennae, but even these two characters were not applicable to all members of Jacobson & Bianki subfamily.

It seems that after Jacobson & Bianki the name *Chrysochraontinae* was not used for the group. The group is restored here to the subfamily rank, but in different scope than in the primary subfamily of Jacobson and Bianki.

Subfamily

Eremogryllinae

(Fig. 67)

Diagnosis: Body small and stout, subcylindrical. Integument finely rugose. Head subconical or subglobular; face, in profile, straight or slightly excurved; fastigium of vertex short, fastigial foveolae present. Antennae filiform or slightly clavate. Dorsum of pronotum flattened and slightly constricted; median and lateral carinae present. Low prosternal convexity present. Mesosternal interspace open, short and wide. Tympanum present. Tegmina and wings fully developed; radial and medial area of male tegmen widened; wing venation not specialized. Hind femora moderately slender; lower lobe shorter than upper one; inner side with row articulated pegs. Lobes of hind knee angular. External apical spine of hind tibia absent. Male cerci short, widened, with incurved, acute apices; supra-anal plate angular with attenuate apex; subgenital plate short, transverse. Ovipositor short, with valves incurved at apices.

Sound-producing mechanism constitute row of articulated pegs on inner side of hind femur and sharp radial vein of tegmen.

PHALLIC COMPLEX: Ectophallus membraneous except sclerotized cingulum; cingulum fully differentiated; zygoma rather wide; apodemes moderately long; rami large; valves of cingulum present. Endophallus strongly sclerotized; penis' sclerites divided on basal and apical valves, connected by slender, rather long flexure; basal valves of penis moderately wide, at proximal ends diverging sideways; gonopore processes present, rather elongated; apical valves short, relatively wide, shorter than valves of cingulum. Epiphallus bridge-shaped, with bridge very narrow in middle with tendency to be divided or divided; ancorae large, incurved not articulated, with bridge; lophi acute tooth-like, up and incurved.

Spermatheca: Variable in its shape.
Karyotype: Unknown.
Type genus: *Eremogryllus* Krauss, 1902.

Besides the type genus, only one genus of this subfamily is known — *Notopleura* Krauss, 1902.

Subfamily *Eremogryllinae* was erected by Dirsh (1956) on the basis of peculiar structure of epiphallus and unusual structure of appendages of the external male genitalia.

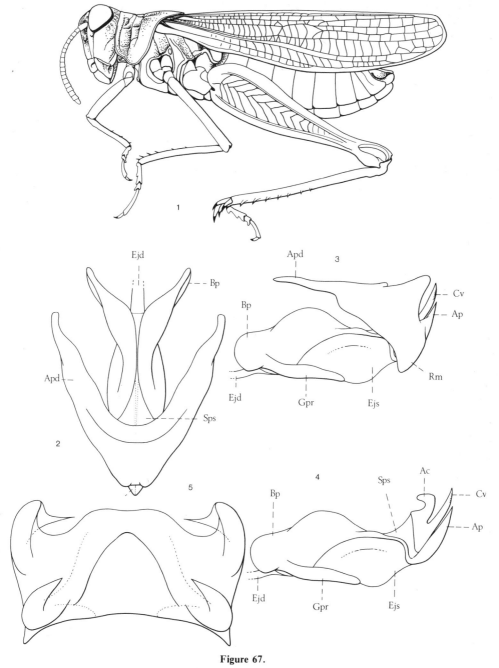

Figure 67.

1, *Eremogryllus hammadae* Krauss, 1902. Male. 2-5, phallic complex. 2, phallic complex, dorsal view (membrane and epiphallus removed). 3, the same, lateral view. 4, endophallus, lateral view. 5, epiphallus.

By the structure of the stridulatory mechanism, the subfamily is linked with subfamily *Gomphocerinae*, otherwise it has very little in common with it. Probably *Eremogryllinae* are relics of the ancestral stock of that subfamily.

Subfamily

Gomphocerinae

(Fig. 68)

Diagnosis: Body from small to medium size, short, cylindrical. Head obtusely conical, short; face, in profile, straight or slightly excurved; fastigium of vertex short, at apex angular or rounded; fastigial foveolae concave or flat, present. Antennae filiform, phyliform or club-like. Dorsum of pronotum flat, subcylindrical or slightly saddle-shaped. Prosternal process absent, rarely low tubercle present. Tegmina and wings fully developed, shortened rarely absent; reticulation sparse; intercalary vein of medial area of tegmen absent. Hind wing sometimes specialized forming widened areas and speculum and thickened costal vein and inflated costal area. Hind femora moderately widened. Knees of hind femur with rounded or obtuse angular lobes.

Stridulatory mechanism represented by serration on inner side of hind femora consisting of row of small pegs forming straight, or regularly undulated, or irregularly shaped line. Pegs articulated to femur by membraneous connection or their bases inclosed into follicles. Sound produced by rubbing row of pegs against sharp, convex radial and medial veins of tegmen.

PHALLIC COMPLEX: Ectophallus membraneous, except strongly sclerotized, well developed cingulum; valves of cingulum present. Endophallus strongly sclerotized; basal valves of penis large and wide, strongly excurved sideways at proximal ends; gonopore processes present; apical valves relatively wide, upcurved; flexure long and relatively wide. Epiphallus bridge-shaped; bridge short and narrow; ancorae short, acute at apices, articulated with ends of bridge; lophi bilobate, rarely monolobate, attached to branches extending from bases of bridge; lateral plates narrow or relatively wide, with angular posterior projections.

Spermatheca: With large, downcurved oval-shaped reservoir and narrow, finger-shaped sometimes twisted and widened one or two diverticula, rarely pear-shaped, without diverticulum. Sometimes apical part divided into two of the same size branches.

Karyotype: $2n\sigma = 17, 22, 23$. In majority of studied cases $2n\sigma = 17$.

Type genus: *Gomphocerus* Thunberg, 1815.

Distribution: Practically all zoogeographical Regions except Australian.

This subfamily is near to the subfamily *Chrysochraontinae*. They have the same type of sound-producing mechanism, the same type of phallic complex and the same shape of spermatheca. The differences are that the antennae of *Gomphocerinae* are mostly filiform while rarely phyliform or with club at apex, in *Chrysochraontinae* they are ensiform or narrow ensiform, and the majority of genera of *Gomphocerinae* have a well developed fastigial foveolae which in *Chrysochraontinae* are absent or are of different type. Also karyotype in *Gomphocerinae* is predominantly $2n\ \sigma = 17$, while in *Chrysocheaontinae* it is predominantly $2n\sigma = 23$. A similar sound-producing mechanism occur only in one more subfamily — *Eremogryllinae*, but latter subfamily differs in the structure of phallic complex.

Gomphocerinae were first established as subfamily by Jacobson and Bianki (1904). They included into the subfamily Palaearctic genera *Gomphocerus*, *Stenobothrus*, *Phlocerus*, *Stauronotus*, *Arcyptera*.

Kirby (1910) however disregarded this subfamily, uniting it with recent *Acridinae* and Jacobson and Bianki *Chrysochraontini*. In that state they remained until 1932, when Tarbinsky used subfamily name *Gomphocerinae* for present *Gomphocerinae* and *Acridinae* (sensu latu).

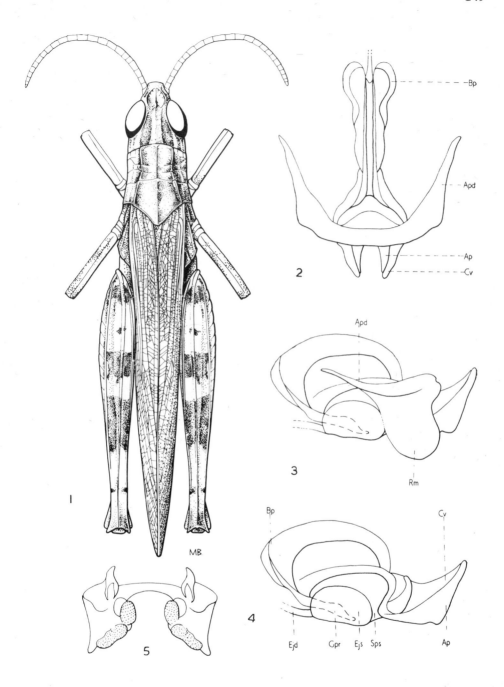

Figure 68.

1, *Dnopherula dorsata*. (I. Bolivar, 1912). Male. 2-5, phallic complex: 2, dorsal view, (epiphallus and ectophallic membrane removed). 3, the same, lateral view. 4, endophallus, lateral view. 5, epiphallus (figs. 2-5, after Hollis, 1966).

Uvarov (1966) resurrected the subfamily *Gomphocerinae* for wide range of genera which are possessing the same type of the stridulatory mechanism and virtually expelling only the genera of the group *Truxales*, for which he reserved subfamily rank *Truxalinae*.

Jago (1969, 1971) used the name and subfamiliar rank *Gomphocerinae* in the sense as Uvarov in 1966.

In present work the subfamily *Gomphocerinae* is recognized as such, but further divided on *Chrysochraontinae* and *Gomphocerinae*.

Subfamily

Truxalinae

(Fig. 69)

Diagnosis: Body strongly elongated, cylindrical, large or medium size. Head strongly elongated, acutely conical; face, in profile, incurved or straight; fastigium of vertex relatively long, at apex rounded or obtuse angular; fastigial foveolae absent. Antennae ensiform. Dorsum of pronotum saddle-shaped, slightly tectiform or flat. Prosternal process or tubercle absent. Tegmina and wings fully developed; reticulation moderately dense; intercalary vein of medial area absent. Hind wings without specialization. Hind femora strongly elongated, narrow. Knees of hind femur with acute lobes.

Stridulatory mechanism represented by serrated ridge on inner side of hind femur and sharp convex radial and medial veins of tegmen.

PHALLIC COMPLEX: Ectophallus membraneous except sclerotized cingulum; zygoma large; apodemes short and narrow; rami moderately wide; valves of cingulum present. Endophallus strongly sclerotized; basal valves of penis relatively small and narrow, slightly protruding sideways at proximal ends; gonopore processes present; apical valves of penis long, straight and relatively wide; flexure moderately narrow and relatively long. Epiphallus bridge-shaped; bridge narrow; ancorae moderately long, acute at apices, articulated with ends of bridge; lophi transverse, bilobate, attached to branches extending from the bases of bridge; lateral plates moderately narrow, with acute posterior projections.

Spermatheca: With large, downcurved, irregularly oval reservoir and finger-shaped diverticulum.
Karyotype: $2n\male = 21, 22, 23$.
Type genus: *Truxalis* Fabricius, 1775.

This subfamily contains only six genera: *Truxalis*, Fabricius 1775; *Truxaloides* Dirsh, 1950; *Acridarachnea* Bolivar, 1908; *Chromotruxalis* Dirsh, 1951; *Xenotruxalis* Dirsh, 1950 and *Oxytruxalis* Dirsh, 1950.

Distribution: Southern part of Palaearctic, Ethiopian and Oriental Regions.

The peculiar shape of body and head in the species of this subfamily is very similar to that in the subfamily *Acridinae* but they differ from each other by the presence of the characteristic stridulatory mechanism in *Truxalinae* and absence of this kind of mechanism in *Acridinae*. (But the latter can produce faint noise by rubbing hind femora against tegmina). The difference in the phallic complex are rather slight but sufficient to consider them on the subfamily level.

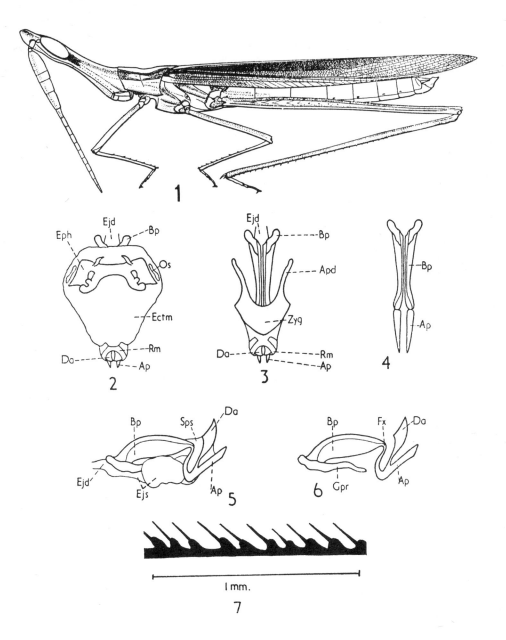

Figure 69.

1, *Truxalis grandis* Klug, 1830. Male. 2-6. Phallic complex of *Truxalis nasutus* (Linnaeus, 1758). 2. Whole phallic complex, from above. 3. The same, but ectophallic membrane and epiphallus removed. 4. Penis from above. 5. As fig. 3, but in profile. 6. Penis in profile. 7. Stridulatory serration on internal side of hind femur.

Genera and species of the *Truxalinae* and *Acridinae* often were confused owing to similar appearance. Both subfamilies were called *Truxalinae* up to 1941 when Roberts changed subfamily name to *Acridinae*.

Tarbinsky (1931) united the present *Acridinae* and *Truxalinae* into one subfamily *Acridinae* and all previously known *Truxalinae* and *Acridinae* he considered as one subfamily which he called *Gomphocerini*. Finally in 1950 Dirsh cleared the interrelation between genera *Truxalis* and *Acrida* and divided subfamily *Acridinae* into two tribes *Truxalini* and *Acridini* on the basis of the presence of the stridulatory mechanism in the former and absence of such form of mechanism in the latter. Later, these tribes were raised to subfamily rank (Dirsh, 1961).

Subfamily

Oedipodinae

(Fig. 70)

Diagnosis: Body from small to large size, comparatively stout, cylindrical or subcylindrical, moderately elongate. Head subglobular to short subconical; face, in profile, straight, excurved or slightly incurved; fastigium of vertex short, subglobular or angular; fastigial foveolae absent or present. Antennae filiform. Dorsum of pronotum tectiform, crest-shaped or saddle-shaped. Prosternal process or tubercle absent. Tegmina and wings fully developed or shortened; reticulation dense; strong and mostly with serrated intercalary vein of medial area. Hind femora short and mostly widened. Knees of hind femur with short, rounded or rarely with angular lobes. Hind tibia sometimes in apical half slightly expanded.

Stridulatory mechanism represented by serrated intercalary vein of medial area of tegmen and by sharp carinula in inner side of hind femur.

PHALLIC COMPLEX: Ectophallus membraneous except sclerotized well developed cingulum; valves of cingulum present. Endophallus strongly sclerotized; basal valves of penis relatively small and narrow, slightly curved sideways at proximal ends; gonopore processes present; apical valves of penis relatively short and wide; flexure moderately short. Epiphallus bridge-shaped, bridge moderately narrow; ancorae short, at apices subacute or obtuse, articulated with ends of bridge; lophi bilobate or less frequently monolobate, attached to inner sides of lateral plates and to branches extending from bases of bridge; lateral plates relatively wide, with subacute or obtuse posterior projections.

Spermatheca: With pear-shaped reservoir and with small diverticulum, sometimes diverticulum is absent.
Karyotype: $2n\sigma = 19, 21, 23, 24$, mostly $2n\sigma = 23$.
Type genus: *Oedipoda* Latreille, 1829.

Distribution: This subfamily is distributed in all zoogeographical Regions.

The most essential character which separates this subfamily from other subfamilies of *Acridoidea* is the presence of the strong and serrated intercalary vein of medial area of a tegmen. However, this character is variable and in some genera, as *Bryodema*, is not serrated at all, but species of this genus can produce sound, when flying, by rubbing strong veins of outer surface corrugated hind wings against strongly convex veins of inner surface of tegmina. In many genera, in females the intercalary vein present but not serrated.

All other differentiating characters *Oedipodinae* share with many other subfamilies of the *Acrididae*. The phallic complex in its main characters is rather similar to that in *Truxalinae* and particularly in *Acridinae*. This indicates comparatively recent divergence of these subfamilies from the common stock.

Oedipodinae were first established as a family by Walker (1870). Since then they were treated as family or subfamily. In 1956 Dirsh proposed to amalgamate this subfamily with subfamily *Acridinae* (sensu lato). Some acridologists, mostly American, accepted this concept, some other acridologists continued to consider *Oedipodinae* as a separate subfamily.

In view of present arrangement of the families and subfamilies, in this work the author decided to restore this group at subfamily rank.

The scope of the subfamily *Oedipodinae* were used by some author differently than is recognized now. Saussure (1884), Jacobson and Bianki (1904) for example included into their family *Oedipodidae* subfamilies *Trinchini* and *Batrachotetrigini* (=*Eremobiini*), which were recognized and transferred in the fifties of this century into family *Pamphagidae*.

The oldest genus of *Oedipodinae* is *Locusta* Linnaeus, 1758 and for the subfamily name it has priority, because the genus *Oedipoda* Latreille, was described in 1829. However the name *Locusta* was used and misused in so many different ways that the present author decided to use the traditional for many years name *Oedipodinae*.

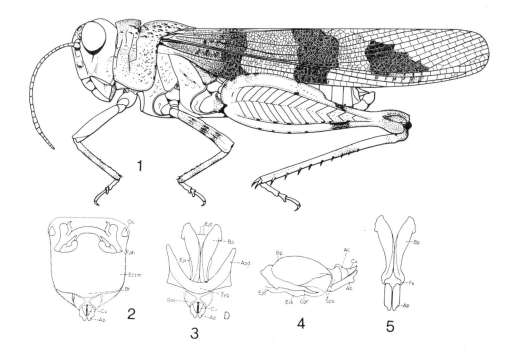

Figure 70.

1, *Oedipoda miniata* (Pallas, 1771). Male. 2-5, phallic complex of *Oedipoda coerulescens* (Linnaeus, 1758). 2, dorsal view. 3, the same, (but membrane and epiphallus removed). 4, the same, lateral view. 5, endophallus, dorsal view.

154

Subfamily

Acridinae

(Fig. 71)

Diagnosis: Body strongly elongated, cylindrical, medium or large size. Head strongly elongated, acutely conical; face, in profile, incurved or straight; fastigium of vertex relatively long, at apex rounded or obtuse angular; fastigial foveolae absent. Antennae ensiform. Dorsum of pronotum flat, slightly tectiform or slightly saddle-shaped. Prosternal process or tubercle absent. Tympanum present. Tegmina and wings fully developed or slightly shortened; weak intercalary vein of medial area of tegmen present. Hind wing with speculum in medial area. Hind femora strongly elongated, narrow. Knees of hind femur with acute lobes.

Stridulatory mechanism present. Sound produced by rubbing of hind wings against tegmina.

PHALLIC COMPLEX: Ectophallus membraneous, except sclerotized, well developed cingulum; valves of cingulum present. Endophallus strongly sclerotized; basal valves of penis relatively small and narrow; slightly curved sideways at proximal ends; gonopore processes present; apical valves of penis relatively long, narrow, almost straight; flexure mostly short. Epiphallus bridge-shaped, bridge narrow; ancorae moderately long, acute at apices, articulated with ends of bridge; lophi transverse,

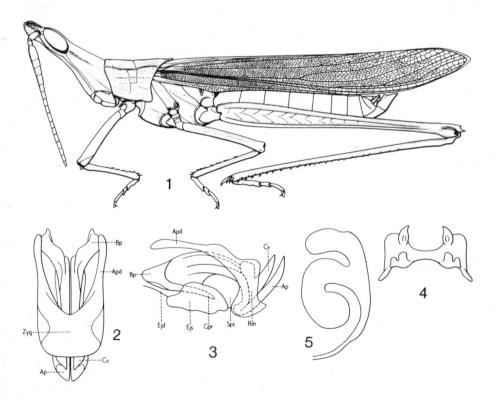

Figure 71.

Acrida subtilis Burr, 1902. Male. 2-4, phallic complex. 2, dorsal view (membrane and epiphallus removed). 3, the same, lateral view. 4, epiphallus. 5, spermatheca.

bilobate, attached to branches extending from the bases of bridge; lateral plates relatively narrow, with acute posterior projections.

Spermatheca: With large, downcurved, irregularly oval reservoir and large, wide diverticulum.

Karyotype: 2n ♂ = 23. (Based on small material).

Type genus: *Acrida* Linnaeus, 1758.

This subfamily contains three genera — *Acrida* Linnaeus, 1758, *Brachyacrida* Dirsh, 1952 and *Chromacrida* Dirsh, 1952.

Distribution: Palaearctic Region (up to 50°N.), Ethiopian Region, Malagassian, Oriental, Austro-Oriental and Australian Regions.

By the peculiar shape of body and head *Acridinae* differs strongly from all other subfamilies of *Acrididae*. There is not intermediate forms which would allow to connect or relate them to the any known subfamily. The exceptions are, however, *Truxalinae* which have strikingly similar external appearance, but differs by the differently specialized stridulatory mechanism and partly by the structure of the phallic complex. In all probability it is example of convergent evolution in both subfamilies.

Tarbinsky (1931) was considering *Acridinae* as a separate subfamily of the family *Acrididae*, containing two genera — *Acrida* and *Acridella* (=*Truxalis*).

Subfamily

Hyalopteryxinae

(Fig. 72)

Diagnosis: Body moderately elongated, cylindrical, medium size. Head acutely conical; face, in profile, slightly incurved; fastigium of vertex moderately long, at apex angular or rounded; fastigial foveolae absent. Antennae ensiform or narrow ensiform. Dorsum of pronotum flat or subcylindrical. Prosternal process absent. Tympanum present. Tegmina and wings fully developed; reticulation of tegmen moderately dense; intercalary vein of medial area absent. Hind wing with highly specialized venation; costal, subcostal, radial and medial veins, in apical half strongly thickened, forming ridge-like projection (stridulatory specialisation); first cubital area strongly widened, forming speculum with sparse, parallel, transverse veinlets. Hind femora moderately widened. Knee of hind femur with lobes rounded or obtuse-angular at apices.

Stridulatory mechanism represented by thickened veins of hind wing and sharp convex radial and second cubital veins in inner side of tegmina. Sound is produced by rubbing tegmina against specialized veins of hind wings.

PHALLIC COMPLEX: Ectophallus membraneous, except well developed sclerotized cingulum; valves of cingulum present. Endophallus strongly sclerotized; basal valves of penis large and wide, at proximal ends curved sideways; gonopore processes present; apical valves of penis moderately slender, upcurved; flexure moderately long and narrow. Epiphallus bridge-shaped; bridge narrow; ancorae moderately long, articulated with ends of bridge, acute at apices; lophi trilobate, with lobes attached to branches of bridge and to inner sides of lateral plates; lateral plates relatively small with comparatively large anterior and small angular posterior projections.

Spermatheca: With downcurved narrow, sack-like main reservoir and comparatively wide and long diverticulum.

Karyotype: 2n ♂ = 23. Based on few observations.

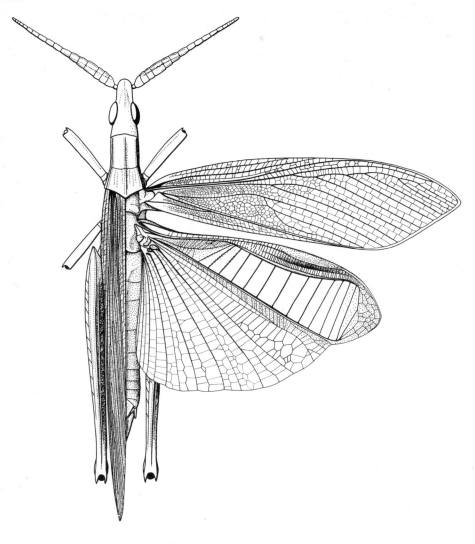

Figure 72.

Hyalopteryx rufipennis (Charpentier, 1845). Male

Type genus: *Hyalopteryx* Charpantier, 1845.

Distribution: South of Nearctic and Neotropical Region.

This subfamily, according to the phallic complex, is nearest to the subfamilies *Phlaeobinae* and *Comphocerinae*. It differs from both by peculiar sound-producing mechanism and by trilobate lophi of epiphallus. However, both these characters, presumably convergently, appear in some unrelated genera of *Gomphocerinae* and *Phlaeobinae*.

The studied genera: *Hyalopteryx* Charpantier, 1845; *Allotruxalis* Rehn, 1944; *Metaleptia* Brunner, 1893; *Thyriptilon* Brunner, 1904; *Paratruxalis* Rehn, 1916.

Subfamily

Gymnobothrinae

(Fig. 73)

Diagnosis: Body moderately elongated small or medium size, subcylindrical. Head conical; face, in profile, straight, slightly excurved; fastigium of vertex short, at apex angular; fastigial foveolae poorly developed in lower position or absent. Antennae filiform or clavate. Dorsum of pronotum flat or subcylindrical. Prosternal tubercle absent. Tympanum present. Tegmina and wings fully developed or shortened. Reticulation dense, intercalary vein of medial area mostly absent, if present then irregular and weakly developed. Hind femora moderately slender. Knee of hind femur with short, rounded lobes.

Stridulatory mechanism not found.

PHALLIC COMPLEX: Ectophallus membraneous except sclerotized, well developed cingulum; valves of cingulum present. Endophallus strongly sclerotized; basal valves of penis large and wide, strongly curved sideways at proximal ends; gonopore processes present; apical valves of penis moderately slender, upcurved; flexure long and narrow. Epiphallus bridge-shaped, bridge narrow; ancorae long, acute at apices, articulated with ends of bridge; lophi relatively long, finger-shaped almost perpendicular to the bridge, with hooked apices; lateral plates very narrow, their posterior projections sub-acute.

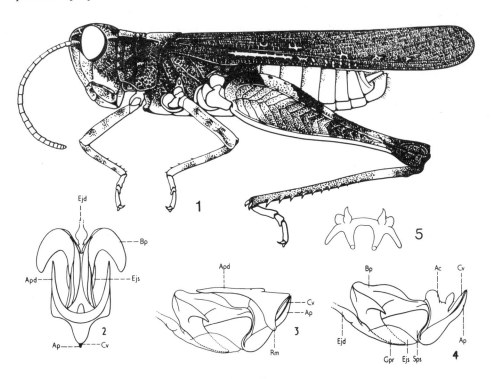

Figure 73.

1, *Gymnobothrus temporalis* (Stal, 1876). Male. 2-5, phallic complex. 2, dorsal view (membrane and epiphallus removed). 3, the same, lateral view. 4, endophallus, lateral view. 5, epiphallus.

Spermatheca: With large, turned down, irregularly oval reservoir and short, narrow diverticulum.

Karyotype: Unknown.

Type genus: *Gymnobothrus* I. Bolivar, 1889. About twelve genera of this subfamily are known.

Distribution: Ethiopian Region, Madagascar.

This subfamily, at present separated from *Acridinae*, have a series of characters which differentiate them clearly. The main character is the shape of epiphallus and the poorly developed fastigial foveolae, placed on the lower part of fastigium of vertex.

Interrelation of *Gymnobothrinae* with other subfamilies of *Acrididae* is still obscure, but some of intermediate genera between this subfamily and subfamily *Gomphocerinae* exist.

Subfamily

Phlaeobinae

(Fig. 74)

Diagnosis: Body moderately or strongly elongate, cylindrical, small or medium size. Head from strongly elongated, acutely conical to short conical; face, in profile, straight or incurved; fastigium of vertex relatively long, at apex rounded or obtuse angular; fastigial foveolae absent. Antennae ensiform or narrow ensiform. Dorsum of pronotum flat or subcylindrical. Prosternal process or tubercle present or absent. Tympanum present. Tegmina and wings from fully developed to strongly shortened; reticulation of tegmen moderately dense; weak, irregular, intercalary vein of medial area present or absent, otherwise tegmina and wings not specialized. Hind femora narrow or moderately widened. Knee of hind femur with lobes angular or rounded at apices.

Stridulatory mechanism not found.

PHALLIC COMPLEX: Ectophallus membraneous, except sclerotized well developed cingulum; valves of cingulum present. Endophallus strongly sclerotized; basal valves of penis large and wide, at proximal ends strongly curved sideways; gonopore processes present; apical valves of penis slender, moderately upcurved; flexure long and comparatively wide. Epiphallus bridge-shaped; bridge narrow; ancorae short, articulated with ends of bridge, acute at apices; lophi wide, monolobate, with lobes from elongated to almost rounded, attached to inner sides of lateral plates and sometimes not clearly differentiated from them, or to extreme bases of bridge; lateral plates moderately narrow, with subacute or obtuse posterior projections.

Spermatheca: With oval downcurved main reservoir and with one finger-shaped diverticulum.

Karyotype: $2n\,\male = 23$. Based on a few observations.

Type genus: *Phlaeobida* Stal, 1860.

Distribution: Ethiopian, Malagassian, Oriental, Austro-Oriental and Australian Regions.

The main characters separating this subfamily from the other subfamilies of *Acrididae* is combination of monolobate lophi attached to inner sides of lateral plates of epiphallus, ensiform antenna, and absence of fastigial foveolae of vertex. All these characters partly are shared with other subfamilies, but combination of them make the subfamily distinct.

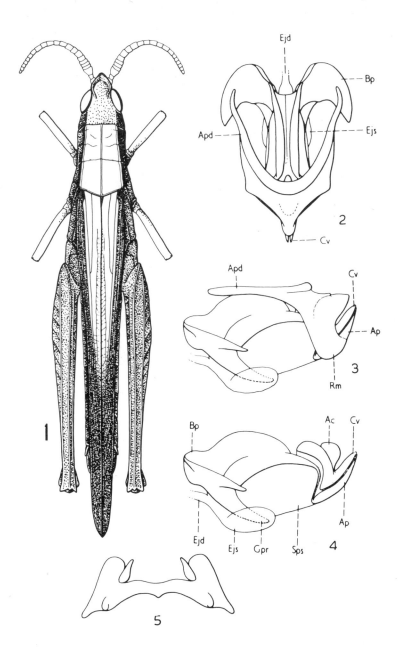

Figure 74.

1, *Duronia chloronota* (Stal, 1876). Male. 2-5, phallic complex of *Plagiacris bimaculata*
Sjostedt, 1931. 2, phallic complex, dorsal view (membrane and epiphallus removed). 3, the
same, lateral view. 4, endophallus, lateral view. 5, epiphallus.

Glossary of terms and symbols

The terminology of the morphological characters used in the taxonomy of the former *Acridoidea* is in a rather confused state. Different authors in different languages and even English-speaking authors often use different terms for the same morphological unit. Taxonomists and anatomists often describe the same thing under different names. The terminology given below, for the characters used in this work, is tentative. Most of the terms listed are used in English papers on the taxonomy of *Acridoidea*, but some are not used in American papers; some of them are borrowed from anatomical works. They are placed in alphabetical order.

Antenna   Paired segmented sensory organ on head.
  Scape   Basal antennal segment attached to head.
  Pedicel   Second antennal segment, between scape and flagellum.
  Flagellum   Distal part of antenna, attached to pedicel.

Main types of antennae:
  Clavate   With apical part of flagellum clubbed or gradually thickened.
  Differentiated   Flagellum clearly divided into basal, medial and apical parts.
  Ensiform   Flagellum compressed, widened in basal part and gradually narrowing towards apex.
  Filiform   Flagellum thin, thread-like. 'Thick filiform' or 'rod-like', as filiform but shorter and thicker.
  Pectinate   Segments of flagellum expanded on one side.
  Phylliform   Flagellum compressed, leaf-like.
  Serrated   With protruding apical angles of the segments forming serration.
Apical fastigial areolae   A pair of areas on apex of fastigium of vertex, at sides of fastigial furrow. Mostly rugose and mostly bounded by carinulae. Occur in *Pyrgomorphidae*.
Apterous   Completely wingless.
Areas of tegmen and wing (see venation).
Arolium   Small, scale-like lobe between tarsal claws.
Basal lobe of hind femur   (see femur).
Brachypterous   With elytra and wings shorter than abdomen, but overlapping or touching each other on dorsally.
Brunner's organ   Small tubercle on lower surface, near base, of hind femur.
Carinae of hind femur   (see femur).
Carinae of pronotum   (see pronotum).
Carinula of fastigium of vertex   Median carinula along fastigium.
Carinula of vertex   Longitudinal carinula in middle of vertex. Often continuous with carinula of fastigium of vertex and occipital carinula.
Cercus   A paired process variably shaped and sized, at base of supra-anal plate and paraprocts.
Claw   One of a pair of claws at apex of last tarsal segment.
Clypeus   Facial sclerite between frons and labrum.
Collar   (see prosternal process).
Coxa   Basal segment of leg, by which leg is attached to body.

160

Elytron = tegmen (pl. Elytra = tegmina). Forewing.

Episternum Small, triangular lobe protruding under lateral lobe of pronotum (part of the pleuron).

External apical spine of hind tibia Spine located apically on external side of hind tibia near spur.

Face Whole anterior part of head visible from the front.

Facial carinae A pair of carinae running between lateral ocelli and clypeus.

Fastigial foveolae A pair of more or less concave foveolae on side of fastigium of vertex, on its anterior margin or below it.

Fastigial furrow A deep thin furrow along middle of apex or whole fastigium of vertex.

Fastigium of vertex Anterior part of vertex. Its base is the shortest line between the eyes, the apex protruding forwards or sloping downwards, merging with frons.

Femur Basal part of leg between coxa and tibia.

Basal lobes of hind femur Two short lobes, upper and lower, forming base of femur.

Carinae of hind femur Upper carina located along upper side of femur. Lower carinae on lower side of femur: *external lower carina* forming lower margin of femur and being a continuation of margin of lower external lobe of knee, and *internal lower carina* being a continuation of margin of internal lower lobe of knee. Between lower carinae tibia fits closely when leg is folded.

Carinulae of hind femur A pair of low carinulae on external and internal sides of hind femur, separating marginal areas from medial; upper carinula and lower.

Areas of hind femur Medial area, external and internal, is central part of femur between upper and lower carinulae. Marginal areas, upper and lower, are marginal parts of femur.

Pattern of hind femur Sculpture of medial area of external side of hind femur.

Fishbone pattern: similar to skeleton of fish.

Feather-like pattern: less convex, resembling feather.

Cellular-pattern: consisting of irregular concavities with irregular ridges between them.

Fenestrae Deep hollows on sides of crest of pronotum, with almost transparent membrane.

Flagellum (see antenna).

Frontal ridge Ridge-like longitudinal convexity on frons between antennae, merging above with fastigium of vertex and below with clypeus, or not reaching clypeus. Sometimes slight or absent; sometimes with lateral carinulae and with sulcus or depression.

Frons Anterior part of face, merging above with fastigium of vertex and below with clypeus on sides with eyes and facial carinae.

Gena Lateral part of head.

Grooves, antennal Pair of deep grooves on sides of frontal ridge, above base of antennae, into which antennae fit closely.

Interocular distance The shortest distance between compound eyes at base of fastigium of vertex.

Knee Apical part of hind femur.

Upper lobe of hind knee Upper lateral part of knee.

Lower lobe of hind knee Lower lateral part of knee.

Crescent Crescent-like area on both sides of hind knee.

Krauss's organ A pair of convexities with ridges or granulose surface at base of first abdominal tergite.

Labrum Upper lip.

Macropterous With tegmina and wings fully developed, exceeding or at least reaching end of abdomen, and overlapping dorsally.

Mandibles Upper jaws.

Mesosternal furcal suture Transverse suture of mesosternum separating it from apical margin of mesosternal interspace; its lateral continuation on either side, fully or partly, forms anterior margins of lateral lobes of mesosternum.

Mesosternal interspace Part of metasternum protruding forwards between mesosternal lobes. May be open, when its posterior end is merging into metasternum, or closed, when mesosternal lobes are connected posteriorly.

Mesosternal lobes A pair of lateral lobes at posterior end of mesosternum, anteriorly separated by mesosternal furcal suture and including between them the mesosternal interspace.

Mesosternum Medial part of sternum, to which medial pair of legs is attached.

Metasturnum Posterior part of sternum, between mesosternum and abdomen.

Metasternal furcal suture Transverse suture of metasternum, separating it from apical margin of metasternal interspace.

Metasternal interspace Part of first abdominal sternit protruding forwards into metasternum. May be open or closed.

Micropterous With greatly reduced tegmina and wings. Elytra not touching each other dorsally.

Occipital carinula Small longitudinal carinula in middle of occiput.

Occiput Posterior part of head merging anteriorly with vertex and usually not having a definite line of separation.

Ocelli Three small, simple eyes, one in middle of frontal ridge and two near inner sides of compound eyes.

Opisthognathous Face sloping backwards with mouth in posterior, ventral position.

Orthognathous Face vertical, with mouth directed downwards.

Ovipositor Six-valves structure (four external and pair internal valves) at end of female abdomen, for digging and depositing the eggs.

Palpi: *Palpi maxillares* — five-segmented appendages of maxillae. *Palpi labiales* — four-segmented appendages of labium.

Paraprocts A pair of lateral lobes, representing parts of eleventh tergite and located on sides of anus, partly below supra-anal plate.

Pedicel (see antenna).

Phallic complex (see p.163).

Pleura Lateral sclerites of thorax.

Presternum Anterior, ridge-like margin of mesosternum.

Prognathous Face sloping forwards, with mouth directed forwards.

Pronotum Dorsal shield of prothorax.

Carinae of pronotum: median carina of pronotum, carina along middle of pronotum; *lateral carinae of pronotum*, a pair of carinae on sides, separating dorsum from lateral lobes.

Dorsum of pronotum Upper part of pronotum either separated from lateral lobes by lateral carinae or gradually merging with them.

Lateral lobes of pronotum Lateral, mainly vertical or sloping parts of pronotum.

Transverse sulci of pronotum There are four transverse sulci on pronotum. The first *(submarginal sulcus)* in most cases present only on lateral lobes and absent on dorsum. The fourth (last) sulcus, is the *basal or posterior sulcus*. Hitherto, almost invariable, the first sulcus has been ignored, so that the second was called the first, and the basal or posterior the third.

Prozona of pronotum Part of pronotum anterior to basal (posterior) sulcus.

Metazona of pronotum Part of pronotum posterior to basal (posterior) sulcus.

Prosternal process A process, tubercle or collar variable in form in middle or at anterior margin of prosternum.

Types of process:

Tubercle-like	Pyramidal
Cylindrical	Cubical
Conical	Bifurcate
Spathulate	Bilobate
Tongue-shaped	Trilobate
Cuneiform	Collar-like
	Beak-like

Remigium Region of tegmen and wing (see venation).

Reticulation (Archedictyon of some authors). Net of veinlets on membrane of tegmen.

Scape (see antenna).

Speculum Expanding, shiny, mostly medial area of hind wing.

Spermatheca Female organ for receiving and storing sperm. Contains *main reservoir* and *spermathecal duct*. Often possess one or several tubular extensions — *diverticula*, or bulges.

Spermatodesma Aggregate of spermatozoids forming a group.

Spermatophore Membraneous reservoir containing sperm for transmitting it to a female.

Spines of hind tibia Two rows of spines on sides of tibia.

Spurs   Two pairs of curved spurs articulated with apex of hind tibia. One pair on external side, one on internal.

Sternum   Ventral part of thorax. (In drawings the meso- and metasternum are denoted as sternum).

Stridulatory serration   Row of teeth or pegs on lower part of inner side of hind femur.

Subgenital plate   In male ninth, in female eighth, abdominal sternite, covering phallic complex in male and genital opening in female.

Subtympanal lobe   A lobe partly covering tympanum, on lower side.

Supra-anal plate   (=epiproct). Eleventh (last) abdominal tergite, covering anus from above.

Tarsus   Three-segmented distal part of leg.

Thorax   Part of body between head and abdomen, to which legs and wings are attached.

Tibia   Part of leg between femur and Tarsus.

Transverse sulci of pronotum   (see pronotum).

Trochanter   Small sclerite between coxa and femur.

Tympanal organ or tympanum   Supposed auditory organ on sides of first abdominal tergite.

Vannus   Region of tegmen and wing (see venation).

Venation   Distribution and pattern of main veins of tegmina and wings and areas between them.

Veins

Precosta (Precostal vein)   A small, secondary, anterior vein on tegmen; often absent.

Costa (Costal vein)   First main vein of tegmen and wing.

Subcosta (Subcostal vein)   Second main vein of tegmen and wing.

Radius (Radial vein)   Third main vein of tegmen and wing. On wing, basal part of radius is fused with next, medial, vein.

Radial sector   Branch of radial vein, which forms several secondary branches denoted R_1, R_2...

Media (Medial vein)   Fourth main vein, in most cases, branches into two; media anterior and media posterior.

Intercalary vein   A short, secondary vein between *M* and *Cu* (in medial area); often stridulatory.

Cubitus (Cubital vein)   Fifth main vein. On tegmen mostly branched into two, on wing unbranched.

Postcubitus (Postcubital vein)   Sixth main vein of tegmen and wing.

Vena dividens (Dividing vein)   Vein separating the remigium from vannus, present on wing only; vannal fold runs along it.

Vannal veins   All veins of vannal part (vannus) of wing and tegmen; numbered 1, 2, 3 etc.

Regions of wings

Remigium   Anterior, unfolded part of tegmen and wing, which includes all veins anterior to vannal fold.

Vannus   Posterior part of tegmen and wing, which includes all vannal veins.

Vannal fold   Line along which remigium and vannus are separated and folded.

Areas

Precostal area   Area anterior to Costa.

Costal area   Area between Costa and Subcosta.

Subcostal area   Area between subcosta and radius.

Radial area   Area between radius and media.

Medial area   Area between media and cubitus.

Cubital area   Area between cubitus and postcubitus.

Postcubital area   Area between postcubitus and vannal vein.

Vannal area   Area posterior to vannal veins; vannal area 1, 2, 3...

Vertex   Upper part of head merging in front with fastigium of vertex and behind with occiput.

Terminology and symbols of the phallic complex (except the epiphallus)

Aa, appendices of penis sclerites (in *Trigonopterygidae*).

Ac, *arch of cingulum*   A small membraneous or sclerotized connection between the zygoma and the valves of the penis or the dorsal wall of the spermatophore sac. It often bears a pair of valves of the cingulum.

Ach, *chamber* A chamber of unknown function formed by the appendices of the penis; probably of ectophallus origin. Found only in the genus *Systella*.

Aed, *aedeagus* The distal complex of the phallus serving as the intromittent organ. Morphologically it represents a complex of elements of the ectophallus and endophallus, though sometimes consisting of either the one or the other. The name can be used only in its functional and not in its morphological sense.

Ap, *apical valves of penis* Distal valves of the penis sclerites in those cases in which the penis is divided into two parts. Their function is to participate in the transmission of the spermatophore to the female genital chamber.

Apd, *apodemes* A pair of usually strongly sclerotized structures, situated dorsally to the endophallus (in the *Trigonopterygidae* ventrally to it); they are a part of the cingulum.

Art, *articulation* Articulated connection between the separated basal and apical valves of the penis.

Ascl A pair of small sclerites situated laterally and internally to the proximal end of the apodemes; found only in *Charilaidae*.

Bf, *basal or dorsal fold* A fold formed by the ectophallic membrane, where, after junction with the cingulum, it extends distally, with the end turned upwards and backwards. Part of the fold sometimes bears sclerotized projections, or sclerotized plates, on the surface. Together with the ventral fold, the basal fold represents an adjustment, which allows the phallic organ to be pushed out and retracted.

Bp, *basal valves of penis* Proximal valves of the penis, when the latter is divided in two parts.

Clf, *cleft* A slit-like ventral opening of the phallotreme where the endophallic membrane is continuous with the ectophallic, or with the rami, or with both.

Cd, *discoidal cingulum* A disc-shaped sclerotization located above the endophallus and derived from the ectophallic membrane.

Cng, *cingulum* A secondary, usually strongly sclerotized structure derived from the ectophallus. It is a capsule or shield-like structure, or may be more complicated, consisting of apodemes, zygoma, rami and valves, sometimes forming the sheath of the endophallus.

Cr Crest on the dorsal surface of the spermatophore sac (occurs in *Euschmidtia* of *Eumastacoidea*).

Cv, *valves of cingulum* A pair of valves projecting from the posterior end of the cingulum either dorsally or ventrally or from the arch of the cingulum. Morphologically derived from the ectophallus. Function: probably auxiliary to the apex of the penis.

Da, *dorsal appendices of penis* A pair of movable sclerites attached to the apical valves of the penis. Their function is probably auxiliary to that of the apical valves. Morphologically they are probably derived from the endophallus.

Dpc A strongly sclerotized dorsal process at the proximal end of the cingulum.

Ect, *ectophallus* External part of the phallic organ consisting mainly of the ectophallic membrane, cingulum and epiphallus, and the oval sclerites when these occur.

Ectm, *ectophallic membrane* Membraneous part of the ectophallus covering the phallic organ. It is continuous with the endophallic membrane.

Ects, *ectophallic sclerites* All sclerites composing the ectophallus.

Ectv, *ectophallic valves* A pair of sclerotized valves representing a continuation of the ectophallic membrane and forming the apex of the aedeagus. The opening of the phallotreme is located between them. (Present in *Eumastacidae*).

Ejd, *ejaculatory duct.*

Ejs, *ejaculatory sac* Distal widening of the ejaculatory duct; situated mostly under the proximal part of the penis. Part of the endophallic sac.

End, *endophallus* Internal part of the phallic organ, consisting of the endophallic sac, penis, and appendices.

Endm, *endophallic membrane* The membraneous part of the endophallus; continuous with the ectophallic membrane.

Endophallic sac Sac of the endophallus consisting of the ejaculatory sac, spermatophore sac and phallotreme.

Ends, *endophallic sclerites* All the sclerites composing the endophallus; derived from the endophallic membrane.

Eph, *epiphallus* Strongly sclerotized sclerite, located on the dorsal side of the phallic organ (except in *Trigonopterygidae*, where the whole phallic complex is in an inverted position).

Et, *endophallic tube* A tube-like structure loosely covering the apical end of the endophallus (in *Proscopia*).

F Fringe-like structures occurring on various parts of the phallus.

Fu Indicates the place of junction of parts of the phallic organ, severed by dissection.

Fuz Junction of the penis with the zygoma.

Fx, *flexure* Flexible sclerotized part of penis connecting its basal and apical valves.

G, *gonopore* Connection between the ejaculatory and spermatophore sacs.

Gpr, *gonopore processes* A pair of ventral processes from the basal valves of the penis.

Gprs, *gonopore sclerites* A pair of small sclerites at the distal end of the ejaculatory duct. (Found only in *Trigonopterygidae*).

M Membraneous parts of the phallic complex.

Mpo Membraneous part of the pouch of the phallus. (In *Trigonopterygidae*).

Mu Sclerotized thickenings, probably for the attachment of muscles.

Op Distal opening of the phallotreme.

P, *penis* In most families, consisting of a pair of elongated, separate (rarely partly fused) sclerites of the endophallus, but in *Eumastacoidea*, it is a single rod-shaped, arched or U-shaped sclerite. Derived from the endophallic sac.

Pal, *pallium* A membraneous continuation of the ectophallic membrane, connected with the walls of the genital chamber. At its distal end the pallium forms a fold partly covering the apex of the phallus, when not erected.

Phallic complex The complex comprising the phallic organ and epiphallus.

Phallic organ or *phallus* The organ concerned with the accumulation of sperm and its transmission to the female genital chamber; it consists of the ecto- and endophallus.

Pht, *phallotreme* Distal extension of the spermatophore sac, with an opening.

Po, *pouch of phallus* A pouch-like structure covering the proximal part of the endophallus. Found only in the genus *Systella* (*Trigonopterygidae*).

Rm, *rami of cingulum* The lateral parts of the cingulum, sclerotized or membraneous, extending from the dorsal part of the cingulum and sometimes connected ventrally. They sometimes form a sheath enveloping the apical part of the penis and are sometimes a membraneous continuation of the ectophallic membrane, and indistinguishable from it.

Scl Sclerites or sclerotized parts of the phallus which do not require special terminology.

Sclph Sclerotized distal lobe of the phallotreme.

Sh, *sheath of penis* The sheath covering the apical part of the penis; it originates from the cingulum and appears as its distal continuation. It may be either strongly sclerotized or membraneous and is sometimes fused with the ventral lobe.

Sps, *spermatophore sac* An extension of the ejaculatory sac, in most cases situated distally and partly dorsally to it. It is connected with the ejaculatory sac by the gonopore, and distally it extends to the phallotreme.

St Sclerotized projection fused with the wall of the ejaculatory sac (only in *Systella*).

Th A thickening of parts of the phallus of unknown significance.

Va Ventral appendices of penis. A pair of valves analogous with the dorsal appendices of the penis, but attached below to the apical valves of it.

Ved Valves of the ejaculatory duct. A pair of valves situated in the distal end of the ejaculatory duct before it widens into the ejaculatory sac.

Vf, *ventral fold* A fold formed by the ectophallic membrane on the ventral side. It is anologous with the basal or dorsal fold (see above) and performs the same function. Apically, this fold sometimes extends into the ventral lobe.

Vinf, *ventral infold* The ventral extension of the ectophallic membrane, sometimes indistinct.

Vlb, *ventral lobe* A lobe formed ventrally by the folding of the ectophallic membrane. It covers the basal part of the apical valves of the penis and the ventral part of the rami of the cingulum. It is sometimes fused below with the rami, and sometimes with the ventral walls of the phallotreme.

Vpc Ventral posterior process of the cingulum.

Zyg, *zygoma* A transverse dorsal part of the cingulum, connecting the apodemes and, in most cases, the cingulum itself with the apical valves of the penis. Morphologically it is not a separate sclerite but only an intermediate part.

Terminology of the parts of the epiphallus and explanation of symbols

A, *ancorae*   A pair of processes of projections from the anterior margin of the dorsal surface of the epiphallus; sometimes they are articulated with the disc.

Ap, *anterior projections*   Projecting anterior ends of the lateral plates.

B, *bridge*   The middle part of the epiphallus connecting the lateral plates, or its lateral parts.

Disc   Central part of the body of the epiphallus in the disc-shaped forms of the latter. In other forms it is a narrow bridge. Sometimes it is divided sagittally into two separate halves.

DA, *dorso-lateral appendices*   Additional lobes connected with the dorso-lateral parts of the anterior end of the epiphallus. Found only in *Pyrgomorphidae*.

L, *lophi*   Processes on or near the posterior end of the epiphallus, sometimes arising from the lateral parts of the bridge, sometimes from its lateral plates. They are very variable in shape.

Lp, *lateral plates*   A pair of symmetrical plates, forming the sides of the epiphallus, sometimes connected with the bridge only by membrane.

Lscl, *lateral sclerite*   Sclerite of unknown significance. (Occurs in *Erucius* of *Eumastacidae*).

Mp, *median projection*   A projection from the middle of the anterior margin of the epiphallus.

Ms, *median slit*   A slit which divides the epiphallus sagittally into two halves, connected only by membrane.

Os, *oval sclerites*   A pair of small sclerites, circular, oval or irregular in form, located laterally to the epiphallus.

Pp, *posterior projections*   Posterior ends of the lateral plates, which sometimes project strongly.

Va, *ventro-lateral appendices*   Irregularly shaped, weakly sclerotized lobes, connected with the ventro-lateral part of the anterior margin of the epiphallus (in the *Charilaidae* only).

Vscl   Ventral sclerite of epiphallus, of unknown significance. (Occurs in *Erucius* of *Eumastacoidea*).

References

AKBAR, S. S., 1968. Systematic history of the family Pyrgomorphidae (Acridoidea: Orthoptera). *S. U. Sci. Res. J.*, *3*(2):121-129.

ANDER, K., 1939. Vergleichend-anatomische und phylogenetische Studien uber die Ensifera (Saltatoria). *Opusc. ent.* Suppl. II, Lund.

BEIER, M., 1955. Orthopteroidea. *In Bronn's Klassen und Ordnungen.* Leipzig.

BERTHOLD, A. A., 1827. *Latreille's naturliche Familien des Thierreichs.* Weimar.

BEY-BIENKO, G. Y. & MISHCHENKO, L. L., 1951. Acridoidea of the Fauna of USSR and neighbouring countries. (In Russian). *Moscow, Akad. Nauk SSSR.* Part 1: 378 pp.

BEY-BIENKO, G. Y., 1962. On the general classification of insects. *Rev. d'Entom. USSR.*, 41, 1: 6-21.

BEY-BIENKO, G. Y., (et al) 1972. Insects and mites, pests of agricultural plants. *I. "Nauka"*, Leningrad.

BLANCHARD, E., 1837. Monographie du genre *Ommexecha* de la famille des Acridiens. *Ann. Soc. ent. Fr.* 5: 603-24.

BLACKITH, R. E. & BLACKITH, R. M., 1968. A numerical taxonomy of orthopteroid insects. *Aust. J. Zool.*, **16**: 111-131.

BLACKITH, R. E., 1972. Morphometrics in acridology: a brief survey. *Acrida*, 1: 7-15.

1973. Clues to the Mesozoic evolution of the *Eumastacidae. "Acrida"*, 3: 5-18.

BOLIVAR, C., 1932. Estudios sobre Eumastacidos. V. Sobre los gencros *Orchetypus* Brunn. *Kirbyita* C. Bol. y *Hemierianthus* Sauss. *Soc. Ent. Fr.*, Livre du Centenaire, 15 juin 1932.

BOLIVAR, I., 1884. Monografia de los Pirgomorfinos. *An. Soc. esp. Hist. nat.* 13: 1-73, 420-500.

1905. Notas sobre los Pirgomorphidos. *Bol. Soc. esp. Hist. nat.* 5: 105-15, 196-217, 278-89, 298-307.

1909. *Pyrgomorphinae. Genera Insect. fasc.* **90**, 40 pp.

1916. *Pamphaginae. Genera Insect.*, **170**: 1-39.

BORDAS, L., 1897. Morphologie des appendices de l'extremite anterieure de l'intestine des Orthopteres. Morphology of the appendages of the extreme anterior of the intestine of Orthoptera. *C. r. hebd. Seanc. Acad. Sci.*, *Paris*, **124**: 376-378.

1897. Classification des Orthopteres d'apres les caracteres tires de l'appareil digestif. Classification of Orthoptera by the characteristics of the digestive apparatus. *C. r. hebd. Seanc. Acad. Sci.*, *Paris*, **124**: 821-823.

BROWN, H. D., 1967. A new Grasshopper allied to *Shelfordites* Karny, 1910 from South West Africa (Orthoptera: Acridoidea). *"Cimbebasia"*, Ser, *A*, **1**, no 1: 1-22.

BRUNNER VON WATTENWYL, C., 1882. Prodromus der europäischen Orthopteren. Leipzig, 1-466.

1893. Revision du systeme des Orthopteres et description des especes rapportees par M. Leonardo Fea de Birmanie. *Ann. Mus. Stor. nat. Genova*, (2), **13**: 1-230.

BRYANTSEVA, I. B., 1953. Peculiarities in the structure of the foregut of Acridids of the subfamily *Egnatiinae* (Orthoptera, Acrididae). *Rev. Ent. U.R.S.S.* 33. Leningrad.

BURMEISTER, H., 1838. *Handbuch der Entomologie, Band 2, 2 Abt.* — Berlin, pp. 591-664, 1,013-1,014.

1840. AUDINET-SERVILLE, Histoire naturelle des Orthopteres — Paris, 1839, verglichen mit H. BURMEISTER, Handbuch der Entomologie, 2 Bd., 2 Abt., 1. Halfte (vulgo Orthoptera). — Berlin, 1938, *Z. Ent.* Leipzig, **2**: 1-82.

BURR, M., 1899. Essai sur les Eumastacides. *An. Soc. esp. Hist. nat.* **28**: 75-112, 253-304, 345-350.

1903. Genera Insectorum. Orthoptera, Fasc. **15**. *Eumastacidae.* 23 pp.

CARBONELL, C. S., & MESA A., 1972. Dos nuevos y especies de acridoideos andinos (Orthoptera). *Anal. Congr. Latinoamericanus Entom.*, **15**, no 1: 95-102.

CHINA, W. E., & MILLER, N. C. E. Check-list and keys to the families and subfamilies of the Hemiptera-Heteroptera. *Bull. Brit. Mus. (Nat. Hist.) Entom.* **8** no 1: 1-45.

CHOPARD, L., 1920. Recherches sur la conformation et le developpement des derniers segments abdominaux chez les Orthopteres. Rennes. Imprimerie Oberthur: 352 pp.

1949. Ordre des Orthopteres. In Grasse, P.P. (Ed.) *Traite de Zoologie* 9. Paris, Masson & Cie: 617-722.

COCKERELL, T. D. A., 1909. Descriptions of Tertiary Insects VII. *Amer. J. Sci. New Haven* (4) XXVIII: 283-286.

1926. Tertiary Insects from Argentina. *Amer. J. Sci. New Haven*, (5) XI: 501-504.

167

168

CRAMPTON, G. C., 1927. The abdominal structures of the Orthopteroid family *Grylloblattidae* and the relationships of the group. *Pan. Pacific Ent.*, 3: 115-135.

DESCAMPS, M., 1964. Revision preliminaire des *Euschmidtiinae* (*Orthoptera — Eumastacidae*). *Mem. Mus. Nat. Hist. Natur. (A)*, 30: 1-321.

1968. Un Acridoide relique des Mascareignes. *Bull. Soc. Ent. France.* 73: 31-35.

1970. Les Eumastacidae de Socotra (Orth.) *Bull. Soc. Ent. Fr.*, 75: 123-134.

1971. Troisieme contribution a l'etude des *Pseudoschmidtiinae*. (Acridomorpha Eumastacidae). *Mem. Mus. Nat. Hist. Nat.*, A, 45: 1-252.

DIRSH, V. M., 1950. Revision of the group *Truxales* (Orthoptera, Acrididae). *Eos.* Madrid, Tomo extraord.: 119-247.

1952. The restoration of the subfamily *Trigonopteryginae* Walker (Orthoptera, Acrididae). *Ann. Mag. Nat. Hist.* (12) 5: 82-84.

1953. *Charilainae*, a new subfamily of Acrididae (Orthoptera). *Ann. Mag. Nat. Hist.*, (12) 6.

1954. *Lathicerinae*, a new subfamily of *Acrididae*, (*Orthoptera*). *Ann. Mag. Nat. Hist.*, (12) 7.

1955. *Tanaoceridae* and *Xyronotidae*: two new families of *Acridoidea* (Orthoptera). *Ann. Mag. Nat. Hist.*, (12) 8.

1956. The phallic complex in *Acridoidea* (*Orthoptera*) in relation to taxonomy. *Trans. R. Ent. Soc. Lond.*, 108, p. 7: 223-356.

1961. A preliminary revision of the families and subfamilies of *Acridoidea* (Orthoptera). *Bull. Brit. Mus. (Nat. Hist.)*, 10, No. 9.

1963. Three new genera and species of the family *Pneumoridae*. *Eos, Madrid*, 39: 177-184

1964. The structure of the phallic complex in the genus *Thericles* (Preliminary report). *Eos, Madrid*, 40: 117-121.

1965. Revision of the family *Pneumoridae* (*Orthoptera: Acridoidea*). *Bull. Brit. Mus. (Nat. Hist.)*, 15, No. 10: 325-396.

1966. Acridoidea of Angola. "*Diamang*" *Publ. Cult. Lisboa*, No. 74: 1-527.

1968. The post-embryonic ontogeny of *Acridomorpha*. *Eos, Madrid*, 43: 413-514.

1970. *Acridoidea* of the Congo (Orthoptera). *Mus. R. Afr. Centre.* (8), No. 182: 1-605.

1973. Genital organs in *Acridomorphoidea* (*Insecta*) as taxonomic character. *Z. f. zool. Syst. u. Evolutionsforschung.* Bd. 11, H. 2: 133-154.

1973. Natural parameters of systematics. *Forma et Functio,* 6: 293-304.

DUPORTE, E. M., 1946. Observations on the morphology of the face in insects. *Jour. Morph.* 79: 371-417.

EADES, D. C., 1961. The tribes and relationships of the *Ommexechinae* (Orthoptera, Acrididae). *Proc. Acad. Nat. Sci. Philad.*, 113, 7: 157-172.

EWER, D. W., 1958. Notes of acridid anatomy. V. The pterothoracic musculature of *Lentula callani* Dirsh, 1956. *J. Ent. Soc. S. Afr.* 21, i.

FABRICIUS, J. C., 1775. Systema entomologiae, pp. 269-293, 826-827.

1777. Genera insectorum, pp. 89-96.

1781. Species insectorum, vol. 1, pp. 340-371.

1793. Entomologia systematica, vol. 2, pp. 1-62.

FISCHER, L. H., 1853. Orthoptera Europea. Leipzig: 1-454.

GERSTACKER, A. 1869. Beitrag zur Insekten-Fauna von Zanzibar. II. *Arch. Naturgesch.* 35 (1): 201-23.

1889. Charakteristik einer Reihe bemerkenswerther Orthopteren. *Mitt. naturw. Ver. Greifswald*, 20: 1-58.

GMELIN, J. F., 1790. Systema Naturae. Part 4.

GRANT, H. J., & RENTZ, D. C., 1967. A biosystematic review of the family *Tanaoceridae* including a comparative study of the proventriculus. *Pan-Pacific Entomologist*, 43, no 1: 65-74.

HANDLIRSCH, A., 1908. *Die fossilen Insecten und die Phylogenie der rezenten Formen.* Leipzig.

HASKELL, P. T., 1961. Insect sounds. Witherby Ltd. London.

HELWIG, E. R., 1958. Cytology and taxonomy. *Bios*, 29, no 2: 58-72.

HENNIG, W., 1969. Die Stammesgeschichte der Insecten. W. Kramer, Frankfurt a. Main: 1-436.

IMMS, A. D., 1970. A general textbook of entomology. Methuen, London: 1-886.

JAGO, N. D., 1971. A review of the *Gomphocerinae* of the world with a key to the genera (Orthoptera, Acrididae). *Proc. Acad. Nat. Sci. Philad.*, 123. 8: 205-343.

JACOBSON, G. & BIANKI, V. L., 1904. *Orthoptera and Pseudoneuroptera of the Russian Empire.* St. Petersburg.

JANNONE, G., 1939. Studio morphologico, anatomico e istologico del *Dociostaurus maroccanus* (Thunb.) nelle sue fasi transiens congregans, gregaria e solitaria. Portici, 1939 - XVIII.

JOHN, B. & HEWITT, G. M., 1968. Patterns and pathways of chromosome evolution within the Orthoptera. *Chromosoma* 25: 40-74.

JOHNSTON, H. B., 1956. Annotated Catalogue of African grasshoppers. *Cambr. Univ. Press.*

KERKUT, G. A., (ed.) *et al.*, 1961. The invertebrata. Cambr. Univ. Press, Cambridge.

KEVAN, D. KEITH, McE., 1952. On the systematic position of two anomalous genera previously placed in the subfamily *Pyrgomorphinae* (Orth., Acrididae). *Ent. mon. Mag.*, 88.

1959. A study of the genus *Chrotogonus* Audinet-Serville, 1839 (*Orthoptera: Acridoidea:* Pyrgomorphidae). *Publ. cult. Co. Diam. Ang.*, 43: 1-246.

1963. A revision of the Desmopterini (Orthoptera: Acridoidea: Pyrgomorphidae). Part 1. Genera other than Desmopterella. *Nova Guinea*, **8**: 197-203.

KEVAN, D. KEITH, McE., & AKBAR, S. S., 1964. The Pyrgomorphidae (Orthoptera: Acridoidea): their systematics, tribal divisions and distribution. *Canad. Entom.*, **96**: 1,505-1,536.

KEVAN, D. KEITH, McE., AKBAR, S. S., & YU-CHEN CHANG, 1970. The concealed copulatory structures of the *Pyrgomorphidae*. *Eos, Madrid*, **45**: 173-228. Part II.

KEVAN, D. KEITH, McE., 1970. A revision of the Desmopterini. (Part II. *Desmopterella* Ramme, 1941). *Pacific Insects*, **12** (3): 543-627.

KIRBY, W. F., 1902. Report on a collection of African *Locustidae* formed by Mr. W. L. Distant chiefly from the Transvaal. *Trans. ent. Soc. Lond.*, 1902: 57-114, 231-41.

1910. A Synonymic Catalogue of the Orthoptera. Vol. 3. *Orthoptera Saltatoria*. Part II. Locustidae vel Acridiidae. London.

LAMARCK, S. B. P. A. DE M. DE, 1815. Histoire naturelle des Animaux sans vertebres, presentant les characteres generaux de ces Animaux. 7 vols.

LAPORTE, F. L. DE CASTERNEAU., 1832. Essai d'une classification systematique de l'ordre des Hemipteres (Heteropteres). *Magasi Zool.* **2**: 52-55.

LATREILLE, P. A., 1793. *Precis des caracteres Generiques des Insectes* Bordeaux.

1802-4. *Histoire naturelle generale et particuliere des Crustacees et des Insectes. Orthoptera, Acrididae*, **3**: 280-84; **12**: 137-64.

1817. Les Crustaces, les Arachnides et les Insectes. Cuvier (G.]C.F.D.). *Le Regne Animal*. Paris.

LIEBERMANN, J., 1943. Generes y especies de nuevos Acridoideos Chilenos. *Rev. Soc. Ent. Argent.*, **2** (5).

LINNAEUS, C., 1758. *Systema Naturae*, 10th ed. **1**: 424-33.

LUBISCHEW, A. A., 1969. Philosophical aspects of taxonomy. *Ann. Rev. Entom.*, **14**.

MASON, J. B., 1954. Number of antennal segments in adult Acrididae (Orthoptera). *Proc. R. ent. Soc. Lond.* (B) **23**, Parts 11-12: 228-238.

1969. The tympanal organ of Acridomorpha. *Eos, Madrid* **44**: 267-355.

MATSUDA, R., 1958. On the origin of the external genitalia of insects. *Ann. Ent. Soc. Amer.*, **51**: 84-94.

1965. Morphology and evolution of the insect head. *Mem. Amer. ent. Inst.*: 1-334.

MAYR, E., LINSLEY, G. E., USINGER, R. L., 1953. Methods and principles of systematic zoology. McGraw-Hill. N.Y.

MAYR, E., 1965. Numerical Phenetics and Taxonomic Theory. *Syst. Zool.*, **14**, No 2: 73-97.

1968. Theory of biological classification. Nature, **220**, No 5167, Nov. 9, 1968: 545.

1969. Principles of Systematic Zoology. New York. McGraw-Hill Co. 428 pp.

McCLUNG, C. E., 1908. Cytology and taxonomy. *Kansas Univ, Bull.*, **4**, no &.

MAKINO, S., 1951. An atlas of chromosomes number in animals. Ames. Iowa.

MISTSHENKO, L. L., 1952. Catantopinae. *Fauna U.S.S.R.* Leningrad.

PICKFORD, R. & GILLOTT, 1971. Insemination in the migratory grasshopper, *Melanoplus sanguinipes* (Fabr.). *Canad. J. Zool.*, **49**, 12: 1583-1588.

RAGGE, D. R., 1955. The Wing-venation of the Orthoptera Saltatoria. *Bull. Brit. Mus. (Nat. Hist.)*. London.

REHN, J. A. G., 1941. On new and previously known species of *Pneumoridae* (Orthoptera: Acridoidea) *Trans. Amer. Ent. Soc.*, **67**: 137-159.

1948. The Acridoid family *Eumastacidae* (Orthoptera). A review of our knowledge of its components, features and systematics, with a suggested new classification of its major groups. *Ibid*. 100.

REHN, J. A. G., & GRANT, H. J., 1958. The phallic complex in the subfamilies of New World Eumastacidae and the family *Tanaoceridae*. *Ibid*. 110.

REHN, J. A. G., 1959. On certain Old World genera of *Teratodini* recently placed in the subfamily Romaleinae. *Not. Nat. Acad. Philad.* No. 317.

1959. An analysis of the tribes of the Romaleinae with special reference to their internal genitalia (Orthoptera: Acrididae). *Trans. Amer. ent. Soc.* **85**.

1959. A review of the Romaleinae (Orthoptera: Acrididae) found in America north of Mexico. *Proc. Acad. Nat. Sci. Philad.* **111**.

REHN, J. A. G., & GRANT, H., 1961. A monograph of the Orthoptera of North America. Monographs of *Acad. Nat. Sci. Philad.*, **12**: 1-255.

ROBERTS, H. R., 1941. A comparative study of the subfamilies of Acrididae (Orthoptera) primarily on the basis of their phallic structures. *Ibid*. 93.

SAEZ, F. A., 1957. An extreme karyotype in an Orthopteran insect. *Amer. Nat.*, **43**: 259-264.

SAUSSURE, H., 1888. Additamenta ad Prodromum Oedipodiorum. *Mem. Soc. Phys.* Geneve, **30** (1).

1899. Orthoptera. In Wissenschaftliche Ergebnisse der Reisen in Madagascar und Ostafrika in den Jahren 1889-95 von Dr. A. Voeltzkow. *Abh. Senckenb. Naturf. Ges.*, **21**: 567-664.

SCUDDER, S. H., 1890. The pretertiary insects of North America. Fossil insects of North America I. p. 455. New York, 1890.

SERVILLE, J. G. A., 1831. Revue methodique des insectes de l'ordre des Orthopteres. *Ann. Sci. Nat. (Zool.)*, **22**: 28-65, 134-62, 262-92. Separate 101 pp. (Dec. 1838). *Histoire naturelle des Insectes*. In Roret, *Collection des Suites a Buffon. Orthopteres*, 776 pp., Paris.

SHAROV, A. G., 1968. Phylogeny of Orthopteroid Insects. Moscow. 1968.

SCHRODER, CH., 1925. *Handbuch der Entomologie*. Jena.

SHVANWITSH, B. N., 1949. *General Entomology* (in Russian). Leningrad.

SLIFER, E. H., 1939-43. The internal genitalia of female Acridinae. *J. Morph*, **65**: 3; **66**: I; **67**: 2; **72**: 2.

1944. Ileal caeca in the *Eumastacidae* (Orthoptera). *Ann. ent. Soc. Amer.* **37**: 4.

SMART, J. 1953. On the wing venation of *Physemacris variolosa* (Linn.) (Insecta: Pneumoridae). *Proc. Zool. Soc. Lond.*, **123** (1): 199-202.

SNODGRASS, R. E., 1935. The abdominal mechanism of a grasshopper. *Smithson. Misc. Coll.* **94**: 6. Washington.

1935. Principles of Insect morphology. McGraw-Hill. N.Y.

1937. The male genitalia of Orthopteroid insects. *Smithson. Misc. Coll.*, **94** (5): 1-107.

1947. The insect cranium and the "epicranial suture" *Smithson. Misc. Coll.*, **107** (7': 1-52.

1957. A revised interpretation of the external reproductive organs of male insects. *Smithson. Misc. Coll.*, **135**, No 6.

SOKAL, R. R., & SNEATH, P. H. A., 1963. Principles of numerical taxonomy. Freeman & Co. Lond.

STAL, C., 1873. Recensio Orthopterorum, **1**: 1-154.

1876. Observations Orthopterologiques. (2) *K. Svenska VetenskAkad. Handl.* **4** (5): 1-58.

1878. Systema Acridiodeorum. *K. svenska VetenskAkad. Handl.* **5** (4): 1-100.

TARBINSKY, S. P., 1931. Review of the palearctic species of the genera *Gomphocerus* Thunb. and *Dasyhippus* Uv. (Acrididae). Izv. Inst. of Pest and Disease Control, **1**: 127-157.

1940. The saltatorian orthopterous Insects of the Azerbaidzhan S.S.R. Leningrad, 1940: 1-245.

THUNBERG, C. P., 1775. *Pneumora* et nytt genus igbland insecterne, uptackt och beskrifvit. *K. Svenska VetenskAkad. Handl.* **36**: 254-60.

1815. Acrydii descriptio. *Nova Acta Soc. Sci. upsal.* **7**: 157-62.

1815. Hemipterorum maxillosorum genera illustrata plurimisque novis speciebus ditata ac descripta. *Mem. Acad. Sci. St. Petersb.* **5**: 211-301.

1824. Grylli Monographia illustrata. *Mem. Acad. Sci. St. Petersb.* **9**: 390-430.

TUXEN, S. L., (edit.), 1970. Taxonomist's glossary of genitalia in insects. Munksgaard: Copenhagen.

UVAROV, B. P., 1943. The Tribe *Thrinchini* of the subfamily *Pamphaginae* and the interrelations of the Acridid subfamilies. *Trans. R. ent. Soc. Lond.* **93**: I.

1948. Recent advances in Acridology. *Anti-Locust Bull.* **1**. London.

UVAROV, B. P., & POPOV, G. B., 1957. The saltatorial Orthoptera of Socotra. *Journ. Linn. Soc. Lond.*, (Zool.) 43.

UVAROV, B. P., 1966. Grasshoppers and locusts. Cambr. Univ. Press. Vol. **1**: 1-475.

UVAROV, B. P., & Dirsh, V. M., 1961. The diagnostic characters, scope and geographical distribution of the subfamily *Romaleinae* (Orthoptera: Acrididae). *Proc. R. Ent. Soc. Lond.*, (B) **30**: 153-160.

WALKER, F., 1870. *Catalogue of the Specimens of Dermaptera Saltatoria in the Collection of the British Museum*, Part III, pp. 485-594; Part IV, pp. 605-801.

1871. *Catalogue of the Specimens of Dermaptera Saltatoria etc.* Supplement, Part V: 49-89.

WALKER, E. M., 1919, 1922. The terminal structures of Orthopteroid insects. *Ann. ent. Soc. Amer.* **12** (4), **15** (I).

WEBER, H., 1938. *Lehrbuch der Entomologie*. Jena.

WESTWOOD, J. O., 1941. *Arcana Entomologica*. London.

WHITE, M. J. D., 1957. Cytogenetics and systematic entomology. *Ann. Rev. Entom.*, **2**: 71-90.

1970. The value of cytology in taxonomic research on Orthoptera. Paper read at Intern. Study Conference on the current and future problems of Acridology.

WILLEMSE, C., 1931. Beschreibung von einigen neuen *Acridoidea* von den Solomon-Inseln. Subfam. Catantopinae, Orthoptera. Jahr. 25, No. 33.

ZEUNER, F. E., 1937. Description of new genera and species of fossil Saltatoria. *Proc. R. Ent. Soc. Lond.*, (B) **6**: 154-159.

1939. Fossil Orthoptera Ensifera. Lond. Brit. Mus. (Nat. Hist.).

1941. The Fossil Acrididae (Orth. Salt.). Part I. Catantopinae. *Ann. Mag. Nat. Hist.*, (11), **8**: 510-22.

1942. The Fossil Acrididae (Orth. Salt.). Part II. *Oedipodinae*. *Ann. Mag. Nat. Hist.* (11), **9**: 128-34.

1942. The Fossil Acrididae (Orth. Salt.). Part III. Acridinae. *Ann. Mag. Nat. Hist.* (11), **9**: 304-14.

1944. The Fossil Acrididae (Orth. Salt.). Part IV. Acrididae incertae sedis and Addendum to Catantopinae. *Ann. Mag. Nat. Hist.* (11), **11**: 359-83.

Alphabetical Index

171